D0859573

Applied Codeology

Navigating the *NEC*® 2017

Applied Codeology, Navigating the *NEC* 2017 is intended to be an educational resource for the user and contains procedures commonly practiced in industry and the trade. Specific procedures vary with each task and must be performed by a qualified person. For maximum safety, always refer to specific manufacturer recommendations, insurance regulations, specific job site and plant procedures, applicable federal, state, and local regulations, and any authority having jurisdiction. The *electrical training ALLIANCE* assumes no responsibility or liability in connection with this material or its use by any individual or organization.

© 2017 *electrical training ALLIANCE*

This material is for the exclusive use by the IBEW-NECA JATCs and programs approved by the *electrical training ALLIANCE*. Possession and/or use by others is strictly prohibited as this proprietary material is for exclusive use by the *electrical training ALLIANCE* and programs approved by *the electrical training ALLIANCE*.

All rights reserved. No part of this material shall be reproduced, stored in a retrieval system, or transmitted by any means whether electronic, mechanical, photocopying, recording, or otherwise without the express written permission of the *electrical training ALLIANCE*.

1 2 3 4 5 6 7 8 9 - 17 - 9 8 7 6 5 4 3

Printed in the United States of America

M54896

Contents

Contents

Contents

Features

For additional information related to QR Codes, visit qr.njatcdb.org Item #1079

Quick Response Codes (QR Codes) create a link between the textbook and the Internet. They can be scanned using Smartphone applications to obtain additional information online. (To access the information without using a Smartphone, visit qr.njatc.org and enter the referenced Item #.)

Facts offer additional information related to the *Codeology* method.

14 Applied Codeology Navigating the *NEC* 2017

FACT

Part I. General applies to all parts within an article. For example, **Part I. General of Article 680 Swimming Pools** applies to **Part II** through **Part VIII.**

referring to the table of contents, Article 110 is divided into five parts:

I. General
II. 1000 Volts, Nominal or Less
III. Over 1000 Volts, Nominal
IV. Tunnel Installations over 1000 Volts, Nominal
V. Manholes and Other Electric Enclosures Intended for Personal Entry, All Voltages

Each part is subdivided into individual sections for clarity, with each section representing either a rule or a part of a rule. Each section is identified by a bold number and has a title. This organizational method allows *Code* users to find the information they are looking for quickly. Sections are sometimes divided further in up to three levels of subdivisions to clarify a requirement. Subdivisions may contain lists, exceptions, and informational notes. First- and second-level subdivisions are always given titles, and third-level subdivisions are permitted to have a title.

The Tables in Chapter 9 are applicable only when referenced in other chapters of the *NEC*. For example, Chapter 9, Table 1 is used for "conduit fill" and is applicable only where referenced in another section (mostly Chapter 3) of the *NEC*. Chapter 9, Table 1 can be used to calculate raceway fill. **Article 358 Electrical Metallic Tubing: Type EMT** is subdivided into three parts. **Part II** is titled "Installation" and includes **Section 358.22**, which refers the user to Table 1 in Chapter 9. This specific reference allows the use of the table.

FACT

The *NEC* structure starts at the Table of Contents and is then broken down into chapters, articles, parts (Roman numerals), sections (XXX.X), and subdivisions (A-B-C, 1-2-3, a-b-c, etc.)

358.22 Number of Conductors. The number of conductors shall not exceed that permitted by the percentage fill specified in Table 1, Chapter 9.

Cables shall be permitted to be installed where such use is not prohibited by the respective cable articles. The number of cables shall not exceed the allowable percentage fill specified in Table 1, Chapter 9.

RULES, EXCEPTIONS, AND INFORMATIONAL NOTES

Understanding how the *NEC* is formatted and structured will prevent the misapplication of the *Code*. Where a *Code* section, part or exception is located is just as important as the information in the respective section, part, or exception.

Application of Rules

The structure of the *NEC* is in a progressive, ladder-type format. It is written in outline format. **See Figure 1-7.** For example, a *Code* rule written as a first-level subdivision may also have second- and third-level subdivisions. Starting at the bottom of the hierarchy, a *Code* rule that exists in a third-level subdivision applies only where the third-level subdivision gives further information under the second-level subdivision. In addition, the second-level subdivision is limited to adding further information to the *Code* rule in the first-level subdivision, which applies only under the section in which it exists. The section applies only in the part of the article it is arranged in. The part applies only to the article in which it exists, which is limited to the chapter in which it is located.

The key to properly applying the rules of the *NEC* is to always apply the rule within the part of the article in which it exists. Without an understanding of the outline form of the *NEC*, a new or inexperienced user might attempt to broadly apply a *Code* rule to areas in other sections to which it may not apply. Using the *Codeology* method, the user will always know in which part of what article and chapter the section exists. This basic information is critical to the proper application of all *NEC* rules.

Exceptions

Exceptions are limited to use only where absolutely necessary. Exceptions are always italicized in the *Code* for quick and easy identification. The *NEC* Code-Making Panels strive to make the *Code* as

Chapter 6 Chapters 5, 6, and 7 of the *NEC*, "Special" 115

Figure 6-5 Hazardous-Rated Raceway Components

Figure 6-5. Equipment that creates sparks when utilized must be connected with a raceway system that is installed for hazardous areas.

2. The likelihood that a flammable or combustible concentration or quantity is present

500.5(B) Class I Locations

Class I locations are those in which flammable gases or flammable liquid-produced vapors are or may be present in the air in quantities sufficient to produce explosive or ignitable mixtures.

(1) Class I, Division 1
- Locations in which ignitable concentrations exist under normal operations
- Locations in which ignitable concentrations may exist due to repair, maintenance, or leaks
- Locations in which ignitable concentrations exist due to processes, breakdown, or faulty equipment

(2) Class I, Division 2
- Locations in which volatiles are handled, processed, or used in closed containers
- Locations in which positive ventilation prevents accumulation of gases/vapors
- Areas adjacent to Class I, Division 1 locations

500.5(C) Class II Locations

Class II locations are those that are hazardous because of the presence of combustible dust. An example of a Class II location is a grain storage bin that produces combustible dust during the movement of the grain. **See Figure 6-6.**

Other Class II hazardous location examples are flour and feed mills; producers of plastics, medicines, and fireworks; producers of starch or candies; spice-grinding plants; sugar plants; and cocoa plants.

(1) Class II, Division 1
- Locations in which ignitable concentrations of combustible dust exist under normal operations
- Where mechanical or machinery failure, repair, maintenance, or leaks could create ignitable concentrations of dust
- Locations in which metal dusts exist, including, but not limited to, aluminum and magnesium

For additional information, visit qr.njatcdb.org Item #1073

Figure 6-6 Storage of Grain Indicates a Hazardous Location

Figure 6-6. Across the heartland of America, farmers utilize grain storage bins that must be hazardous rated. It makes no difference whether it is a couple of storage bins or a large distribution center; they are all considered hazardous.

***Code* Excerpts** are "ripped" from *NFPA 70®* or other sources.

Headers and **Subheaders** organize information within the chapter.

Figures, including photographs and artwork, clearly illustrate concepts from the text.

Features

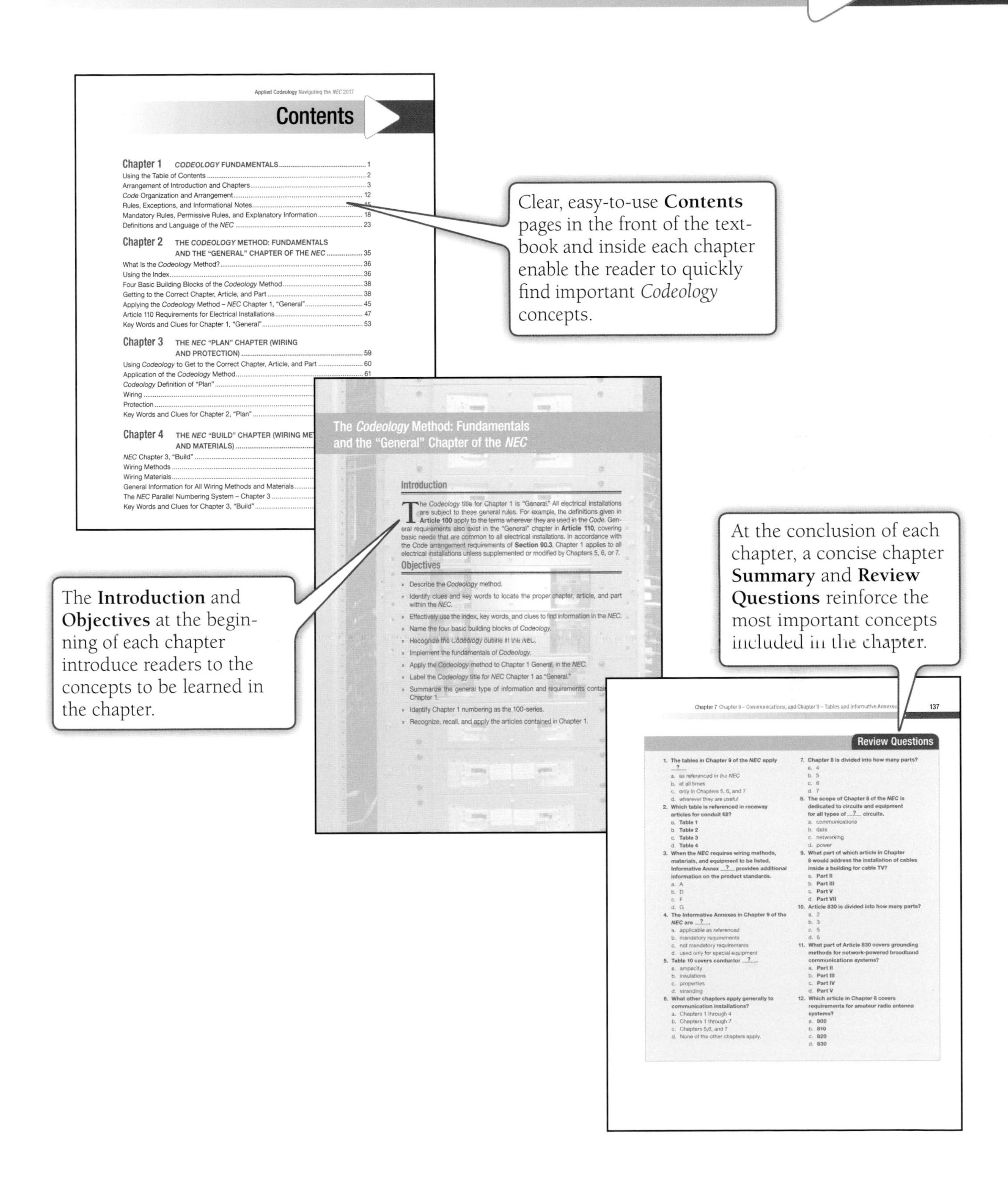
Contents
Clear, easy-to-use **Contents** pages in the front of the textbook and inside each chapter enable the reader to quickly find important *Codeology* concepts.
The *Codeology* Method: Fundamentals and the "General" Chapter of the *NEC*
Introduction
Objectives
The **Introduction** and **Objectives** at the beginning of each chapter introduce readers to the concepts to be learned in the chapter.
At the conclusion of each chapter, a concise chapter **Summary** and **Review Questions** reinforce the most important concepts included in the chapter.
Review Questions

Introduction

Apprentices, Journeymen Electrical Workers, electrical engineers, electrical maintenance workers, electrical inspectors, and numerous other Electrical Workers will use the *NFPA 70: National Electrical Code® (NEC)* on a daily basis. Their livelihood depends upon their ability to properly design, install, inspect, and maintain electrical systems in accordance with the *NEC*.

The *NEC* can be overwhelming to new apprentices in the electrical industry. In fact, some Electrical Workers complete an apprenticeship or an electrical engineering degree without confidence in their ability to quickly find references within the *NEC*. This can be very frustrating since they must use the *NEC* on a daily basis to complete their work. Add to this the fact that the *NEC* changes every three years and this frustration continues to grow.

An experienced tradesperson can differentiate between members of different trades just by looking at the tools they carry. Electrical Workers use hand tools, the most identifiable being side-cutting pliers. To an apprentice, a pair of side-cutters seems clumsy, but necessary. After a few years of field experience, those same side-cutters seem to become an extension of the hand, capable of many tasks; all of which are performed quickly and efficiently. The *NEC* is a "tool" which Electrical Workers use on a daily basis. Like any other tool, the Electrical Worker must learn to use it properly in order to successfully complete the task required. When and only when Electrical Workers understand how to use the *NEC*, they will be able to successfully, efficiently, and safely complete the tasks required of them.

About This Book

The *electrical training ALLIANCE Applied Codeology* textbook is designed to help the student understand how to use the *National Electrical Code® (NEC)*. The word "Codeology" is derived from two words: "code," referring to the *National Electrical Code*, and "ology," meaning an academic field of study. Thus, *Applied Codeology* is the academic study of the *NEC* both in the classroom and on the job. The original concept of *Codeology* was developed in the 1960s by Ronald O'Riley, a prominent author of electrical training materials, who used it in his training program. Over the years the *electrical training ALLIANCE* secured the rights to publish the book and it has been a fundamental part of the *electrical training ALLIANCE* apprenticeship training program for over four decades.

Since the *NEC* is used both nationally and internationally, it is vital that the organization, structure, and language be as clear and explicit as possible. The *NEC* is written using a patterned method established by the National Fire Protection Association (NFPA). The Correlating Committee is charged by NFPA to implement the *Code* process. The 2015 *National Electrical Code Style Manual* is available on the NFPA website to direct those interested in the *NEC* process by providing a standard for organization, arrangement, language, and methods used to write code.

The *Code* is written in a very organized and methodical way. With an understanding of the organizational structure of the *NEC*, the *Code* user can efficiently and confidently maneuver through the articles and sections easily, finding the information needed to answer a question or complete a task. The objective of *Applied Codeology* is to provide the user with basic skills in understanding how the *Code* book is organized and help students and users build confidence in their ability to quickly find pertinent information within the *NEC.* In a classroom setting, *Applied Codeology* textbook content may lead to discussion of specific *Code* rules. Keep in mind that the intent of *Codeology* is to be a tool used to develop confidence in finding information in the *NEC*.

Using the *Applied Codeology* method, students will discover and describe the fundamental structure of the *NEC*. In addition, *Applied Codeology* will introduce methods and strategy that the user can apply immediately to improve the skills needed to find information in the parts, sections, and subdivisions used in *Code* sections.

Acknowledgments

Special thanks to those who shared their expertise and encouragement during the updating of this book. Updating a book to the current *NEC* and closely following the publishing of the *NEC* in a timely manner is a challenge. The book would not look like it does or be readable without the patience of the *electrical training ALLIANCE's* curriculum department, which oversaw editing, formatting, image revision, and other non-electrical content. Where any illustrations needed changes, Topher Edwards with the Portland, Oregon Electrical JATC shared his skilled eye for detail, which guided any updates.

One more note: as the *NEC* has been updated over the years, this book has been revised, updated, and illustrated by many other experts in the field. Their marks remain on the book and continue to contribute today.

QR Credits
National Fire Protection Association (NFPA)
Legrand
Arcat
3M
Eaton
Thomas & Betts, a member of the ABB Group
Hubbell Power Systems, Inc.
Rockwell Automation
Emerson Industrial Automation
SunWize
Pentair Engineered Electrical and Fastening Solutions
CDI Torque Products

About the Author

ABOUT THE SUBJECT MATTER EXPERT UPDATING THIS EDITION

John McCamish has been involved in the electrical industry for over 25 years. After graduating from the University of Oregon, he pursued a career in the electrical industry, first as an apprentice, then as a Journeyman, and then as a foreman. After working for over 10 years in the field, he has instructed apprentice and Journey-level electricians in various aspects of the trade, including Fire Alarm, Instrumentation, Code and Calculations, and Theory among others.

McCamish currently serves his community as a member of a local code appeals board and serves nationally on the Fundamentals Technical Committee for *NFPA 72: The National Fire Alarm and Signaling Code* and on Code-Making Panel 2 of the *National Electrical Code*.

Codeology Fundamentals

Introduction

A qualified person working in the electrical industry must have the proper skills, knowledge, and attitude to design, install, and maintain safe electrical installations. The *electrical training ALLIANCE* provides standardized electrical training to accomplish this goal. The *National Electrical Code*® (*NEC*), published by the National Fire Protection Association (NFPA), is a critical reference tool for anyone working in the electrical industry. Users of the *NEC* can benefit from an organized system designed to address the layout and content of the *Code*. The *Applied Codeology: Navigating the National Electrical Code* textbook method provides the electrical professional a systematic way to find information in the vast catalog of articles contained in the *NEC.*

There are four basic items that will make *Codeology* a successful tool using the *NEC*. The building blocks are:

1. The Contents pages of the *NEC*
2. Section 90.3 *Code* Arrangement
3. The structure of the *NEC*
4. Definitions—the language of the *NEC*

The organization of the *NEC*, the *Code* as written in outline form, definitions, rules, and language are all essential for understanding the application of the *Codeology* method.

Basic principles should be repeated time and time again until they become part of the natural thought process when working with the *NEC.* The *NEC* structure is governed and outlined by the *National Electrical Code Style Manual*. See Appendix A and Appendix B for a discussion on this important document, which is also available for download from the NFPA website. When reviewing any type of installation requirements, the user must understand important elements of the language and the appropriate definitions used in the *Code*. Article 100 contains definitions used in two or more sections of the *NEC*. Several articles have definitions that apply to the sections within that article; these definitions are written within the second section **(XXX.02)** of these articles. Each definition contains language necessary for using the *Code* successfully.

Chapter 1

Objectives

- » Identify articles and parts and highlight the *Code* book for using the *Codeology* method.
- » Name the basic building blocks used in the *Codeology* method.
- » Quickly locate specific articles and parts in the *Code* using the table of contents.
- » Describe the layout of the *Code* and state which chapters apply to general installations and which apply to special installations.
- » Relate the function and topical area installation covered by each chapter in the *Code*.
- » Diagram the outline format used for articles, sections, and subsections.
- » Identify rules, exceptions, and informational notes and explain how each of these is used differently in the *NEC*.
- » Apply and recognize the difference between mandatory, permissive, and explanatory language in the *NEC*.
- » Understand the language of the *Code* and specific definitions as they are used in articles and sections.

Table of Contents

FACT

The Table of Contents is always the starting point when using the *NEC*.

FACT

The nine chapters of the *NEC* logically separate installation requirements to aid the *Code* user in quickly locating needed information.

USING THE TABLE OF CONTENTS

The table of contents (TOC) in the *NEC* is the key to the *Codeology* method and the starting point for all *Code* inquiries. When a need arises for information or to find an answer to a question from the *NEC*, the first step is to use the TOC to reach the right chapter, article, and part. Thus it is essential that all users of the *NEC* be familiar with the TOC and its layout. **See Figure 1-1.**

The TOC comprises ten separate sections or chapters. The first is the introduction to the *NEC*, **Article 90**. Each of the remaining nine chapters covers a broad area of the *Code*. These chapters are subdivided into articles and parts. The introduction (**Article 90**) and each of the nine chapters must also be covered in depth.

Article 90 Introduction
Chapter 1 General
Chapter 2 Wiring and Protection
Chapter 3 Wiring Methods and Materials
Chapter 4 Equipment for General Use
Chapter 5 Special Occupancies
Chapter 6 Special Equipment

Figure 1-1 The *NEC* Table of Contents

Contents

Reprinted with permission from NFPA 70-2017, *National Electrical Code*®, Copyright© 2016, National Fire Protection Association, Quincy, MA 02169. This reprinted material is not the complete and official position of the NFPA on the referenced subject, which is represented only by the standard in its entirety.

Figure 1-1. The table of contents is always the starting place for finding information in the NEC.

Chapter 7 Special Conditions
Chapter 8 Communications Systems
Chapter 9 Tables and Informative Annexes

NFPA 70: National Electrical Code® is revised every three years to address new technologies and public safety issues. As each new edition becomes available, the TOC is expanded and changed to accommodate new sections and parts that must be reviewed by the *Code* user.

ARRANGEMENT OF INTRODUCTION AND CHAPTERS

The first major subdivision of the *NEC* is **Article 90 Introduction**. It provides the basic information and installation requirements necessary to properly apply to the rest of the document.

Article 90 Introduction

Article 90 provides the ground rules upon which the rest of the *NEC* is based. In laying the ground rules, the following sections of **Article 90** establish the format of the *Code*.

Section 90.1 Purpose
Section 90.2 Scope
Section 90.3 Code Arrangement

Section 90.3 Code Arrangement
One of the most important basic facts about the structure of the *NEC* is that the TOC provides a straightforward, ordered design for the proper application of each successive chapter. **Article 90 Introduction** details this arrangement in **Section 90.3** by illustrating the division of the *NEC* into the introduction and nine chapters. **See Figure 1-2.**

As required in **Section 90.3**, Chapters 1 through 4 apply generally in all electrical installations. They contain the basic requirements for all electrical installations, from a single-family dwelling unit to a petroleum refinery or a hospital.

FACT

The NJATC provides a concise overview of *NEC* changes and additions to the most recent edition of the *NEC* in the *Significant Changes to the NEC* book.

Figure 1-2 Chapter Arrangement of the *NEC*

Article 90	Introduction
Chapters 1 through 4 apply GENERALLY to ALL electrical installations.	
Chapter 1	General
Chapter 2	Wiring and Protection
Chapter 3	Wiring Methods and Materials
Chapter 4	Equipment for General Use
Chapters 5, 6, and 7 SUPPLEMENT or MODIFY Chapters 1 through 7.	
Chapter 5	Special Occupancies
Chapter 6	Special Equipment
Chapter 7	Special Conditions
Chapters 1 through 7 DO NOT apply to Chapter 8 unless there is a specific reference in Chapter 8 referring to another chapter.	
Chapter 8	Communications Systems
The tables in Chapter 9 apply as referenced elsewhere in the *NEC*.	
Chapter 9	Tables
Informative Annexes A through J are for informational purposes only and are not mandatory.	
Chapter 9	Informative Annexes A through J

Figure 1-2. Section 90.3 describes the chapter arrangement of the NEC.

Marking Your *Code* Book

Finding information quickly in the *NEC* is important to success as a student and worker. Marking and annotating the *Code* book can be very helpful in finding information. The *Codeology* method starts with the table of contents. Highlighting, marking, and making notes in the table of contents is one of the basic tasks required for using the *Codeology* method. The following suggestions will make any copy of the *NEC* easier to use and will enhance the reader's ability to quickly and accurately find the information by using the *Codeology* method.

Highlighting can be very useful for *Code* users. In fact, certain portions of the *Code* book have already been highlighted by the publisher. Gray highlighting is used for each article title. In the text body of a section, gray highlighting is used to indicate any *Code* changes from a previous edition of the *NEC*. The most important rule for highlighting is to highlight only where absolutely necessary. It is counterproductive to highlight an entire page of the *Code* book.

The most productive use of highlighting is to start with the articles and parts. The articles of the *NEC* are already outlined in gray, so highlighting is not necessary (but is optional and will make the article title stand out.) A green highlighter can be used to box in the gray highlight of the article titles (see the image of **Article 110** below). It is important to highlight the parts in each article. Orange is used to highlight the parts. This color is not used for any other highlighting in the *Code* book, so it will allow the user to instantly identify the parts in each section (see the image of **Article 250** below). Color-coding the parts is extremely important as it allows the user to find a part and move to another part in an article to quickly find answers. While green (optional) is used to highlight the articles, and orange to highlight the parts, yellow is used to highlight important information in a section, but only where necessary.

ARTICLE 110
Requirements for Electrical Installations

I. General

110.1 Scope. This article covers general requirements for the examination and approval, installation and use, access to and spaces about electrical conductors and equipment; enclosures intended for personnel entry; and tunnel installations.

Articles are highlighted in gray from the publisher in the Code book. The user may enhance the highlight with additional green highlighting.

ARTICLE 250
Grounding and Bonding

I. General

250.1 Scope. This article covers general requirements for grounding and bonding of electrical installations, and the specific requirements in (1) through (6).

(1) Systems, circuits, and equipment required, permitted, or not permitted to be grounded
(2) Circuit conductor to be grounded on grounded systems
(3) Location of grounding connections
(4) Types and sizes of grounding and bonding conductors and electrodes
(5) Methods of grounding and bonding
(6) Conditions under which guards, isolation, or insulation may be substituted for grounding

Parts of an article are highlighted in orange.

The following exercise is mandatory for all *Codeology* students.

Open the *NEC* to the table of contents. The table of contents is extremely useful in finding information in the *NEC*. Highlighting it will make it easier to use. Notice that most articles in the table of contents have a high voltage section. Highlight these parts in the table of contents with an orange highlighter. If a question arises regarding higher voltages, these entries will be easier to find.

Chapter 3 of the *NEC* covers cable assemblies, raceways and support methods among other topics. Each of the wiring methods typically have their own acronym (NMC, IGS, MI, etc.)

Marking Your *Code* Book—*continued*

If the user is not familiar with each acronym and it is used in a question, highlighting the acronyms in the table of contents with a yellow highlighter will aid the user.

Continue with Article 90. Highlight the article title in green (if desired), and then move to Article 100 and highlight the parts in orange

ARTICLE 100 Introduction

I. General

II. Over 1000 Volts, Nominal

Move on to Article 110 and locate the page numbers for the article and parts. Find the parts and highlight them in orange.

ARTICLE 110 Requirements for Electrical Installations

I. General

II. 1000 Volts, Nominal, or Less

III. Over 1000 Volts, Nominal

IV. Tunnel Installations over 1000 Volts, Nominal

V. Manholes and Other Electric Enclosures Intended for Personnel Entry, All Voltages

Continue through the table of contents, finding the pages for each part in the *Code* and highlighting each article (optional) and part with green and orange. Marking and highlighting will allow the user to quickly find information. When applying the *Codeology* method, it is imperative to know the article and part in which each *Code* section is located to apply the installation requirement. Another effective practice is to use yellow highlighting or use a red pen or pencil to underline text, subdivisions, key words, list items, and any other information that may be useful in the future. **See Article 300 below.** This method helps to find previously studied material and is extremely useful when revisiting specific areas of the *Code*.

ARTICLE 300
General Requirements for Wiring Methods and Materials

Part I. General Requirements

300.1 Scope.

(A) All Wiring Installations. This article covers general requirements for wiring methods and materials for all wiring installations unless modified by other articles in Chapter 3.

(B) Integral Parts of Equipment. The provisions of this article

Exception: Individual conductors shall be permitted where installed as separate overhead conductors in accordance with 225.6.

(B) Conductors of the Same Circuit. All conductors of the same circuit and, where used, the grounded conductor and all equipment grounding conductors and bonding conductors shall be contained within the same raceway, auxiliary gutter, cable tray, cablebus assembly, trench, cable, or cord, unless otherwise permitted in accordance with 300.3(B)(1) through (B)(4).

(1) Paralleled Installations. Conductors shall be permitted to be run in parallel in accordance with the provisions of 310.10(H). The requirement to run all circuit conductors within the same raceway, auxiliary gutter, cable tray, trench, cable, or cord shall apply separately to each portion of the

Important material should be highlighted in yellow or underlined. It is important not to highlight the entire page of the Code.

FACT

Gray highlighting found in the text body of an article outlines any *Code* changes from a previous edition of the *NEC*. A black bullet in the left margin indicates a section or text removed from the *NEC* from the last edition.

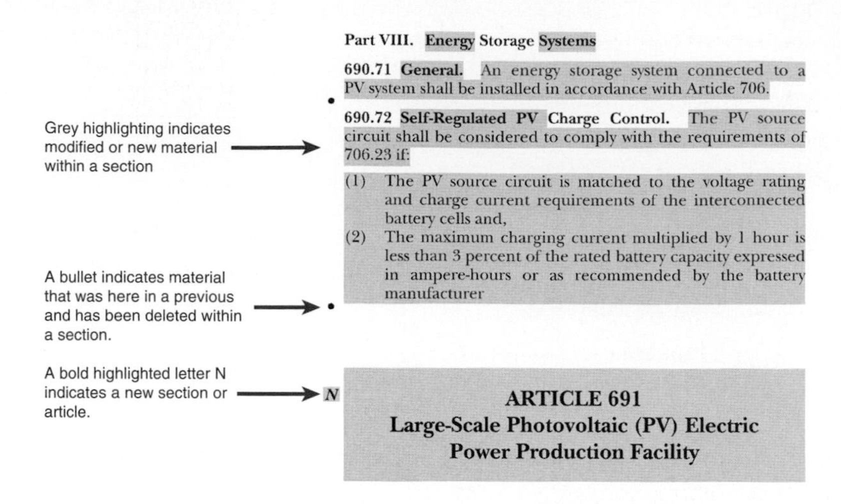

A comprehensive understanding of these four chapters is imperative because they are the backbone for all electrical installations. These sections set forth enforcement, rules, Informative Annexes, formal interpretations, safety examinations, and wiring planning.

Chapters 5, 6, and 7 are for special occupancies, equipment, and conditions that are supplemental, modify, or amend the general rules of Chapters 1 through 7. Chapter 8 is not subject to the requirements of Chapters 1 through 7 except where the requirements are specifically referenced in Chapter 8. Chapter 9 includes tables that are referenced as requirements in the *Code*. The Annex material in the back of the book provides information related to installations and is not a mandatory requirement of the *Code*.

FACT

The scope of a chapter or article is the range covered by an activity, subject, topic, or a range of application. Each article in the *NEC* has a section in XXX.1 titled "Scope" that gives an overview of what will be covered in that article. See **110.1** as an example of the scope of an article.

Chapter 1 General

Although the title of Chapter 1 is simply "General," its scope is general information and rules for electrical installations, and it pertains to all chapters of the *NEC*. It is essential to become familiar with the information in Chapter 1 as a foundation for basic installation requirements used in all buildings and structures.

Chapter 1 Consists of 2 Articles:

100 Definitions
110 Requirements for Electrical Installations

Chapter 2 Wiring and Protection

The scope of Chapter 2 is information and rules on wiring and protection of electrical installations. **Article 200** through **Article 230** address wiring and include information related to grounded conductors, calculations for conductor size, branch circuits, feeders, and services. **Article 240** through **Article 285** address protection and include information related to the use of overcurrent protection, grounding, bonding, surge arresters, and surge protection devices (SPDs.) **See Figure 1-3**.

Chapter 2 Consists of 10 Articles:

200 Use and Identification of Grounded Conductors
210 Branch Circuits
215 Feeders

220 Branch-Circuit, Feeder, and Service Load Calculations
225 Outside Branch Circuits and Feeders
230 Services
240 Overcurrent Protection
250 Grounding and Bonding
280 Surge Arresters, Over 1 kV
285 Surge-Protective Devices (SPDs), 1 kV or Less

Chapter 3 Wiring Methods and Materials

Chapter 3 provides information on all permitted methods and materials to supply an electrical installation. This chapter details requirements for installation from the service point to termination at the last outlet in the electrical distribution system. The articles in Chapter 3 provide general information for all wiring methods, types of cable assemblies, types of raceways, cabinets, cutout boxes, meter socket enclosures, boxes, conduit bodies, fittings, and more.

FACT

Chapter 3 allows for the use of three basic wiring methods, raceways, cables, and open wiring on insulators in limited use. It is important to know that cords, such as SJ, SOW, and so on, are not considered a wiring method and have restrictive installations rules. These are covered in Chapter 4 – see Article 400.

Chapter 3 can be divided into four general topical categories. The first addresses the general rules for all wiring methods, which include the general installation requirements of **Article 300**, conductors in **Article 310**, and cabinets, enclosures, and boxes in **Article 312** and **Article 314.** Second is the types of cable assembles in **Article 320** through **Article 340.** Notice that the cable assemblies are in alphabetical order. Third are raceways in **Article 342** through **Article 390**, starting with the metal conduits, both rigid and flexible; then the nonmetallic conduits; and then tubing. The raceways covered in **Articles 342** to **362** tend to be wiring methods with a circular cross section, while **Articles 366** to **390** are mostly wiring methods that have a rectangular cross section. Finally, everything else from **392** to **399** are open wiring or support methods.

Figure 1-3 Wiring and Protection

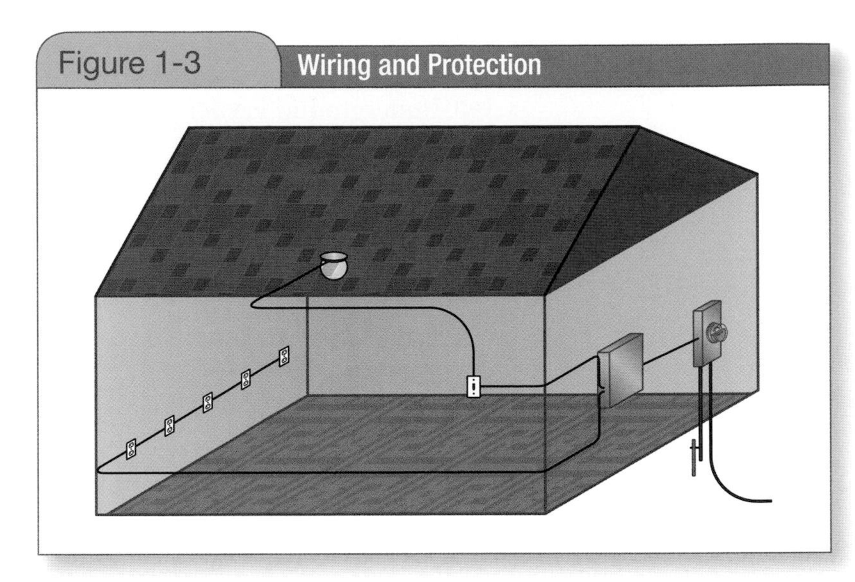

Figure 1-3. Chapter 2, Wiring and Protection, covers the feeders, branch circuits, services, and grounding for most buildings.

FACT

Article 300 is one of the fundamental articles in the *NEC.* It covers wiring requirements for all electrical installations.

__Chapter 3 Consists of 45 Articles:__

300 General Requirements for Wiring Methods and Materials
310 Conductors for General Wiring
312 Cabinets, Cutout Boxes, and Meter Socket Enclosures
314 Outlet, Device, Pull, and Junction Boxes; Conduit Bodies; Fittings; and Handhole Enclosures
320 Armored Cable: Type AC
322 Flat Cable Assemblies: Type FC
324 Flat Conductor Cable: Type FCC
326 Integrated Gas Spacer Cable: Type IGS
328 Medium Voltage Cable: Type MV
330 Metal-Clad Cable: Type MC
332 Mineral-Insulated, Metal-Sheathed Cable: Type MI
334 Nonmetallic-Sheathed Cable: Types NM, NMC, and NMS
336 Power and Control Tray Cable: Type TC

FACT

Cables are generally flexible and come from the manufacturer with conductors pre-installed. Raceways shall be installed as complete systems between outlet, junction, or splicing points prior to the installation of conductors.

338 Service-Entrance Cable: Types SE and USE
340 Underground Feeder and Branch-Circuit Cable: Type UF
342 Intermediate Metal Conduit: Type IMC
344 Rigid Metal Conduit: Type RMC
348 Flexible Metal Conduit: Type FMC
350 Liquidtight Flexible Metal Conduit: Type LFMC
352 Rigid Polyvinyl Chloride Conduit: Type PVC
353 High Density Polyethylene Conduit: Type HDPE Conduit
354 Nonmetallic Underground Conduit with Conductors: Type NUCC
355 Reinforced Thermosetting Resin Conduit: Type RTRC
356 Liquidtight Flexible Nonmetallic Conduit: Type LFNC
358 Electrical Metallic Tubing: Type EMT
360 Flexible Metallic Tubing: Type FMT
362 Electrical Nonmetallic Tubing: Type ENT
366 Auxiliary Gutters
368 Busways
370 Cablebus
372 Cellular Concrete Floor Raceways
374 Cellular Metal Floor Raceways
376 Metal Wireways
378 Nonmetallic Wireways
380 Multioutlet Assembly
382 Nonmetallic Extensions
384 Strut-Type Channel Raceway
386 Surface Metal Raceways
388 Surface Nonmetallic Raceways
390 Underfloor Raceways
392 Cable Trays
393 Low Voltage Suspended Ceiling Power Distribution Systems
394 Concealed Knob-and-Tube Wiring
396 Messenger-Supported Wiring
398 Open Wiring on Insulators
399 Outdoor Overhead Conductors over 1000 Volts

Each article in Chapter 3 has standardized or "parallel" section numbers. As with all articles in the *Code,* the XXX.2 subsection is used for definitions. Chapter 3 uses a similar pattern with other subdivision numbers. **See Figure 1-4.**

Chapter 4 Equipment for General Use

The scope of Chapter 4 is information and rules on equipment for general use in electrical installations. Note that Chapter 3 addresses wiring methods and materials, whereas Chapter 4 addresses all electrical equipment necessary for utilization, control, generation, and transformation of

Figure 1-4 **Chapter 3 Parallel Numbering System**

Parallel Numbering Sections in Chapter 3 Articles		
Article 320 Armored Cable Type AC	**Article 330 Metal Clad Cable Type MC**	**Article 334 Nonmettalic Sheathed Cable Types NM, NMC, NMS**
- 320.2 Definitions	- 330.2 Definitions	- 334.2 Definitions
- 320.10 Uses permitted	- 330.10 Uses permitted	- 334.10 Uses permitted
- 320.12 Uses not permitted	- 330.12 Uses not permitted	- 334.12 Uses not permitted
- 320.24 Bending Radius	- 330.24 Bending Radius	- 334.24 Bending Radius
- 320.30 Securing and Supporting	- 330.30 Securing and Supporting	- 334.30 Securing and Supporting
- 320.40 Boxes and Fitting	- 330.40 Boxes and Fitting	- 334.40 Boxes and Fitting

Figure 1-4. Where possible, similar subjects with the same purposes (for example, XXX.30 Securing and Supporting) are used in section numbers, and part numbers where possible, for the same purposes within articles covering similar subjects.

Making Notes in the *Code* Book

As *Code* studies progress to topic-specific courses like overcurrent protection or grounding and bonding, take the time to write brief, neat, notes in pencil in the book. The last few pages of the *NEC* are blank and can be used for such notes. If the users plans to take a qualifying exam for a state or other electrical license, it may be useful to know that some jurisdictions do not allow written notes in the *NEC* book used for the exam. Check with each jurisdiction and see if this rule applies for each area. A test proctor who finds notes in a *Code* book may require the user to obtain a "clean" copy of the *NEC* for the test. Furthermore, each state or jurisdiction can modify the *NEC* for their region. It can be useful to underline with another color or to use a vertical highlight mark in the margin next to the area of the *NEC* that is modified by a local area.

Code tabs are an extremely effective tool for quickly finding frequently used articles and sections of the *NEC*. Many organizations make *Code* tabs, so it is worthwhile to look at several versions. When applying *Code* tabs, be sure to take the time to apply them properly. Read and follow the installation instructions carefully.

electrical energy in an electrical installation. Chapter 4 includes requirements for equipment cords/cables, switches, receptacles, panelboards, generators, transformers, appliances, motors, and other utilization equipment.

Chapter 4 Consists of 22 Articles:

400 Flexible Cords and Cables
402 Fixture Wires
404 Switches
406 Receptacles, Cord Connectors, and Attachment Plugs (Caps)
408 Switchboards, Switchgear, and Panelboards
409 Industrial Control Panels
410 Luminaires, Lampholders, and Lamps
411 Low-Voltage Lighting
422 Appliances
424 Fixed Electric Space-Heating Equipment
425 Fixed Resistance and Electrode Industrial Process Heating Equipment
426 Fixed Outdoor Electric Deicing and Snow-Melting Equipment
427 Fixed Electric Heating Equipment for Pipelines and Vessels
430 Motors, Motor Circuits, and Controllers
440 Air-Conditioning and Refrigerating Equipment
445 Generators
450 Transformers and Transformer Vaults (Including Secondary Ties)
455 Phase Converters
460 Capacitors
470 Resistors and Reactors
480 Storage Batteries
490 Equipment Over 1000 Volts, Nominal

Chapter 5 Special Occupancies

The scope of this chapter is modifications and/or supplemental information and rules for electrical installations in special occupancies. As stated in **Section 90.3**, this chapter includes "occupancy-specific" information supplementing or modifying the first seven chapters for special occupancies such as hazardous locations, health care facilities, places of assembly, carnivals, agricultural buildings, mobile homes, RVs, and marinas.

Chapter 5 Consists of 28 Articles:

500 Hazardous (Classified) Locations, Classes I, II, and III, Divisions 1 and 2
501 Class I Locations
502 Class II Locations
503 Class III Locations
504 Intrinsically Safe Systems
505 Zone 0, 1, and 2 Locations
506 Zone 20, 21, and 22 Locations for Combustible Dusts, or Ignitable Fibers/Flyings
510 Hazardous (Classified) Locations - Specific
511 Commercial Garages, Repair and Storage
513 Aircraft Hangars
514 Motor Fuel Dispensing Facilities
515 Bulk Storage Plants
516 Spray Application, Dipping, Coating, and Printing Processes Using Flammable Combustible Materials
517 Health Care Facilities
518 Assembly Occupancies
520 Theatres, Audience Areas of Motion Picture and Television Studios, Performance Areas, and Similar Locations
522 Control Systems for Permanent Amusement Attractions
525 Carnivals, Circuses, Fairs, and Similar Events
530 Motion Picture and Television Studios and Similar Locations
540 Motion Picture Projection Rooms
545 Manufactured Buildings
547 Agricultural Buildings
550 Mobile Homes, Manufactured Homes, and Mobile Home Parks
551 Recreational Vehicles and Recreational Vehicle Parks
552 Park Trailers
553 Floating Buildings
555 Marinas and Marinas, Boatyards, and Commercial and Noncommercial Docking Facilities
590 Temporary Installations

Chapter 6 Special Equipment

The scope of Chapter 6 is modifications and/or supplemental information for electrical installations containing special equipment. This chapter includes "equipment-specific" information supplementing or modifying the first seven chapters in regards to special equipment, such as electric signs, welders, x-ray equipment, swimming pools, solar photovoltaic systems, fuel cells, and fire pumps.

Chapter 6 Consists of 27 Articles:

600 Electric Signs and Outline Lighting
604 Manufactured Wiring Systems
605 Office Furnishings
610 Cranes and Hoists
620 Elevators, Dumbwaiters, Escalators, Moving Walks, Platform Lifts, and Stairway Chairlifts
625 Electric Vehicle Charging System
626 Electrified Truck Parking Space
630 Electric Welders
640 Audio Signal Processing, Amplification, and Reproduction Equipment
645 Information Technology Equipment
646 Modular Data Centers
647 Sensitive Electronic Equipment
650 Pipe Organs
660 X-Ray Equipment
665 Induction and Dielectric Heating Equipment
668 Electrolytic Cells
669 Electroplating
670 Industrial Machinery
675 Electrically Driven or Controlled Irrigation Machines
680 Swimming Pools, Fountains, and Similar Installations
682 Natural and Artificially Made Bodies of Water
685 Integrated Electrical Systems
690 Solar Photovoltaic (PV) Systems

691 Large-Scale Photovoltaic (PV) Electric Power Production Facility
692 Fuel Cell Systems
694 Wind Electric Systems
695 Fire Pumps

Chapter 7 Special Conditions

The scope of Chapter 7 is modifications and/or supplemental information for electrical installations under special conditions. This chapter includes "condition-specific" information supplementing or modifying the first seven chapters for special conditions, such as emergency systems; legally required standby systems; Class 1, 2, and 3 systems; and fire alarm systems.

Chapter 7 Consists of 15 Articles:

700 Emergency Systems
701 Legally Required Standby Systems
702 Optional Standby Systems
705 Interconnected Electric Power Production Sources
706 Energy Storage Systems
708 Critical Operations Power Systems (COPS)
710 Stand-Alone Systems
712 Direct Current Microgrids
720 Circuits and Equipment Operating at Less Than 50 Volts
725 Class 1, Class 2, and Class 3 Remote-Control, Signaling, and Power -Limited Circuits
727 Instrumentation Tray Cable: Type ITC
728 Fire-Resistive Cable Systems
750 Energy Management Systems
760 Fire Alarm Systems
770 Optical Fiber Cables

Chapter 8 Communication Systems

Chapter 8 covers communications systems and is not subject to the requirements of Chapters 1 through 7 (per **Section 90.3**) except where the requirements are specifically referenced in Chapter 8. This means that all of the articles listed in Chapter 8 stand alone and are not subject to the rules in the rest of the *NEC* unless a Chapter 8 article specifically references a requirement elsewhere in the *Code*. Chapter 8 includes specific information for communications systems, such as communications circuits, radio equipment, television equipment, CATV, and broadband systems that are both network powered and premises powered broadband systems.

Chapter 8 Consists of 5 Articles:

800 Communications Circuits
810 Radio and Television Equipment
820 Community Antenna Television and Radio Distribution Systems
830 Network-Powered Broadband Communications Systems
840 Premises-Powered Broadband Communications Systems

Chapter 9 Tables and Informative Annexes

Chapter 9 contains tables that are referenced throughout the *NEC*. Per **Section 90.3**, Chapter 9 tables are applicable as referenced and are part of the requirements of the *NEC*.

Informative Annexes, which follow Chapter 9, are not part of the requirements of the *NEC*, but are included for informational purposes.

Chapter 9 Consists of Tables and Informative Annexes:

Table 1 Percent of Cross Section of Conduit and Tubing for Conductors
Table 2 Radius of Conduit and Tubing Bends
Table 4 Dimensions and Percent Area of Conduit and Tubing (Areas of Conduit or Tubing for the Combinations of Wires Permitted in Table 1, Chapter 9)
Table 5 Dimensions of Insulated Conductors and Fixture Wires

Table 5A Compact Copper and Aluminum Building Wire Nominal Dimensions and Areas
Table 8 Conductor Properties
Table 9 Alternating-Current Resistance and Reactance for 600-Volt Cables, 3-Phase, 60 Hz, 75°C (167°F) - Three Single Conductors in Conduit
Table 10 Conductor Stranding
Table 11(A) Class 2 & Class 3 Alternating-Current Power Source Limitations
Table 11(B) Class 2 & Class 3 Direct-Current Power Source Limitations
Table 12(A) PLFA Alternating-Current Power Source Limitations
Table 12(B) PLFA Direct-Current Power Source Limitations
Informative Annex A Product Safety Standards
Informative Annex B Application Information for Ampacity Calculation
Informative Annex C Conduit and Tubing Fill Tables for Conductors and Fixture Wires of the Same Size
Informative Annex D Examples
Informative Annex E Types of Construction
Informative Annex F Availability and Reliability for Critical Operations Power Systems; and Development and Implementation of Functional Performance Tests (FPTs) for Critical Operations Power Systems
Informative Annex G Supervisory Control and Data Acquisition (SCADA)
Informative Annex H Administration and Enforcement
Informative Annex I Recommended Tightening Torque Tables from UL Standard 486A-B
Informative Annex J ADA Standards for Accessible Design

For additional information, visit qr.njatcdb.org Item #1062

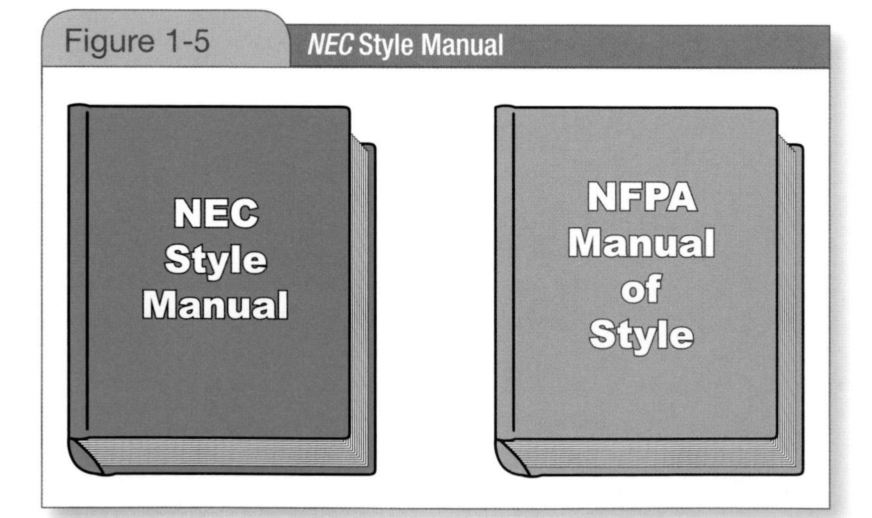

Figure 1-5. Most of the NFPA codes use rules from the NFPA Manual of Style. *The* NEC (NFPA 70) *uses a set of rules specific to the* NEC *called the* NEC Style Manual.

CODE ORGANIZATION AND ARRANGEMENT

New users of the *NEC* may question why the *Code* is written the way it is. Sometimes it can be a hard book to understand, and it requires knowledge of electrical theory, language, and equipment used in the electrical trade. In **90.1(A)**, the *Code* states that it is not intended as a design specification or an instruction manual for untrained persons. However, it does not take long for the user to learn that the *NEC* is a well-organized book, addressing electrical installation standards located in most public and private premises wiring. The foundation of the *Code* is established by rules and regulations governing the structure, organization, and writing style of the *NEC* through a guidebook published by the NFPA called the *NEC Style Manual*.

The *NEC Style Manual* is a practical writing guide developed for use by *Code*-making panels and editors with the intent to make the *Code* as clear as possible. The *NEC Style Manual* shows the document structure and numbering, use of terms and definitions, and how to refer to other NFPA publications. The *NEC Style Manual* is available online at www.nfpa.org. Another important document, not to be confused with the *NEC Style Manual,* is the *NFPA Manual of Style.* Whereas the *NEC Style Manual* is used to develop the *NEC* only, the *NFPA Manual of Style* is used for developing all 300 NFPA Codes and Standards. **See Figure 1-5**.

NEC Organized in Outline Format

The TOC shows that the *NEC* is separated into nine chapters and an introduction. **Article 90** introduces the *NEC* and provides essential information for electrical installations. The first section, **90.1**, describes the purpose of the *Code* as providing practical safeguarding for people and property from the hazards that may arise from the use of electricity. After discussing what is covered and what is not covered by the *NEC* in **90.2, 90.3** provides the arrangement and application requirements.

Articles

Articles are chapter subdivisions and cover specific topics such as grounding, metal conduit systems, cables, motors, transformers, and so on. Each article is given a specific title, and articles that are sufficiently large are sometimes subdivided into parts and into subcategories called *sections*. For example, Chapter 1 is titled "General" because the information is of a general nature and applies to all electrical installations. The two articles in Chapter 1 are titled **Article 100 Definitions** and **Article 110 Requirements for Electrical Installations**. **See Figure 1-6.**

An article that contains a large amount of information, or one for which it is necessary to logically group requirements, can be divided into parts and sections that correspond to logical groupings of information. Parts are designated by Roman numerals and have titles. Parts are subdivided into sections. For example,

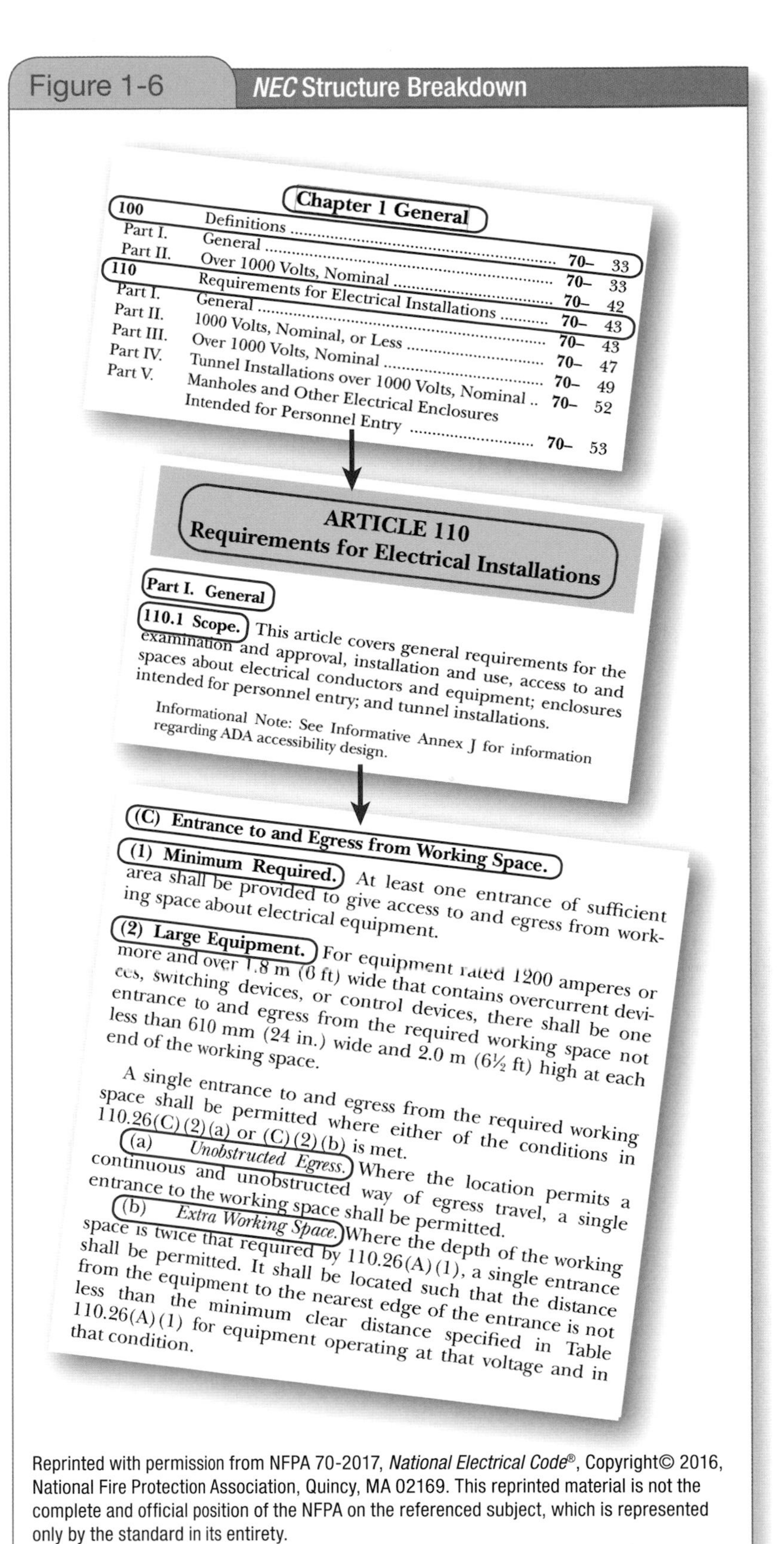

Reprinted with permission from NFPA 70-2017, *National Electrical Code*®, Copyright© 2016, National Fire Protection Association, Quincy, MA 02169. This reprinted material is not the complete and official position of the NFPA on the referenced subject, which is represented only by the standard in its entirety.

Figure 1-6. The NEC *structure starts at the table of contents, and is then broken down into chapters, articles, parts (Roman numerals), sections (XXX.X), and subdivisions (A-B-C, 1-2-3, a-b-c, etc.).*

FACT

Part I. General applies to all parts within an article. For example, **Part I. General** of **Article 680 Swimming Pools** applies to **Part II** through **Part VIII.**

referring to the table of contents, Article 110 is divided into five parts:

I. **General**
II. **1000 Volts, Nominal or Less**
III. **Over 1000 Volts, Nominal**
IV. **Tunnel Installations over 1000 Volts, Nominal**
V. **Manholes and Other Electric Enclosures Intended for Personal Entry, All Voltages**

Each part is subdivided into individual sections for clarity, with each section representing either a rule or a part of a rule. Each section is identified by a bold number and has a title. This organizational method allows *Code* users to find the information they are looking for quickly. Sections are sometimes divided further in up to three levels of subdivisions to clarify a requirement. Subdivisions may contain lists, exceptions, and informational notes. First- and second-level subdivisions are always given titles, and third-level subdivisions are permitted to have a title.

The Tables in Chapter 9 are applicable only when referenced in other chapters of the *NEC*. For example, Chapter 9, Table 1 is used for "conduit fill" and is applicable only where referenced in another section (mostly Chapter 3) of the *NEC*. Chapter 9, Table 1 can be used to calculate raceway fill. **Article 358 Electrical Metallic Tubing: Type EMT** is subdivided into three parts. **Part II** is titled "Installation" and includes **Section 358.22**, which refers the user to Table 1 in Chapter 9. This specific reference allows the use of the table.

FACT

The *NEC* structure starts at the Table of Contents and is then broken down into chapters, articles, parts (Roman numerals), sections (XXX.X), and subdivisions (A-B-C, 1-2-3, a-b-c, etc.)

358.22 Number of Conductors. The number of conductors shall not exceed that permitted by the percentage fill specified in Table 1, Chapter 9.

Cables shall be permitted to be installed where such use is not prohibited by the respective cable articles. The number of cables shall not exceed the allowable percentage fill specified in Table 1, Chapter 9.

RULES, EXCEPTIONS, AND INFORMATIONAL NOTES

Understanding how the *NEC* is formatted and structured will prevent the misapplication of the *Code*. Where a *Code* section, part or exception is located is just as important as the information in the respective section, part, or exception.

Application of Rules

The structure of the *NEC* is in a progressive, ladder-type format. It is written in outline format. **See Figure 1-7.** For example, a *Code* rule written as a first-level subdivision may also have second- and third-level subdivisions. Starting at the bottom of the hierarchy, a *Code* rule that exists in a third-level subdivision applies only when the third-level subdivision gives further information under the second-level subdivision. In addition, the second-level subdivision is limited to adding further information to the *Code* rule in the first-level subdivision, which applies only under the section in which it exists. The section applies only in the part of the article it is arranged in. The part applies only in the article in which it exists, which is limited to the chapter in which it is located.

The key to properly applying the rules of the *NEC* is to always apply the rule within the part of the article in which it exists. Without an understanding of the outline form of the *NEC*, a new or inexperienced user might attempt to broadly apply a *Code* rule to areas in other sections to which it may not apply. Using the *Codeology* method, the user will always know in which part of what article and chapter the section exists. This basic information is crucial to the proper application of all *NEC* rules.

Exceptions

Exceptions are limited to use only where absolutely necessary. Exceptions are always italicized in the *Code* for quick and easy identification. The *NEC Code-Making* Panels strive to make the *Code* as

Figure 1-7 *NEC* Application of Rules

Section	Applies only within the scope of the part of the article in which it is located.
First-Level Subdivision	Applies only within the scope of the section in which it exists.
Second-Level Subdivision	Applies only within the scope of the first-level subdivision in which it exists.
Third-Level Subdivision	Applies only within the scope of the second-level subdivision in which it exists.
List Items	Applies only in the section or subdivision in which they exist.
Exceptions	Applies only to the section, subdivision, or list item under which they exist.
Informational Notes	Used for informational purposes only and are designed to aid the user in the application of the rule(s) under which they exist.

Figure 1-7. Application of rules in the NEC *is in a progressive, ladder-type format.*

user-friendly as possible. Over time, most exceptions have been eliminated and the *Code* rules changed to positive text, making them easier to read and apply. Exceptions are used in the following cases:

1. The *Code* rule in which the exception is applied is modified or supplemented elsewhere in the *NEC*. This modification or supplement will be qualified to very specific locations, equipment, conditions, or wiring methods or uses. Examples of this type of exception include the following:
   210.8(A)(3) Exception
   314.28 Exception
   450.3(B) Exception
2. Existing conditions may require alternative methods or modification of the rule. Examples of this type of exception include the following:
   110.26(A)(3) Exception No.1
   300.20(A) Exception No.1
3. Exceptions are written to allow specific variations from the general rule to explain and clarify the intent and scope of the rule. Examples of this type of exception follow:
   110.26(E)(1)(a) Exception
   210.19(A)(3) Exception No.1

Where an exception to a rule exists, the exception will immediately follow the main rule where it applies. When possible, the Technical Committees (*Code*-making panels) use positive language, usually written in permissive format like "shall be permitted" or "shall not be required," within a given section instead of using an exception to the *Code* rule.

When an exception is made to a section that contains list items, the exception will clearly indicate the items within the list to which it applies. **See Figure 1-8.** Where exceptions are used, they truly are exceptions to the rule beneath which they appear.

Informational Notes

Informational notes are explanatory material and are not an enforceable part of the *NEC* as required in **90.5(C)**. As such, these notes are variable and include, but are not limited to, the following types:

1. Informational
2. Referential (referencing other sections or areas of the *NEC* or other codes)
3. Design
4. Suggestions
5. Examples

FACT

There are two types of Exceptions in the *NEC*:

1. **Mandatory,** where "shall" and "shall not" direct the action.
2. **Permissive,** where "shall be permitted" states that the action is acceptable, but not required.

Figure 1-8 Organizational Numbering of Exceptions in the *NEC*

210.8 Ground-Fault Circuit-Interrupter Protection for Personnel. Ground-fault circuit-interrupter protection for personnel shall be provided as required in 210.8(A) through (E). The ground-fault circuit interrupter shall be installed in a readily accessible location.

Informational Note No. 1: See 215.9 for ground-fault circuit-interrupter protection for personnel on feeders.

Informational Note No. 2: See 422.5(A) for GFCI requirements for appliances.

For the purposes of this section, when determining distance from receptacles the distance shall be measured as the shortest path the cord of an appliance connected to the receptacle would follow without piercing a floor, wall, ceiling, or fixed barrier, or passing through a door, doorway, or window.

(A) Dwelling Units. All 125-volt, single-phase, 15- and 20-ampere receptacles installed in the locations specified in 210.8(A)(1) through (10) shall have ground-fault circuit-interrupter protection for personnel.

(1) Bathrooms
(2) Garages, and also accessory buildings that have a floor located at or below grade level not intended as habitable rooms and limited to storage areas, work areas, and areas of similar use
(3) Outdoors

→ *Exception to (3): Receptacles that are not readily accessible and are supplied by a branch circuit dedicated to electric snow-melting, deicing, or pipeline and vessel heating equipment shall be permitted to be installed in accordance with 426.28 or 427.22, as applicable.*

(4) Crawl spaces — at or below grade level
(5) Unfinished portions or areas of the basement not intended as habitable rooms

→ *Exception to (5): A receptacle supplying only a permanently installed fire alarm or burglar alarm system shall not be required to have ground-fault circuit-interrupter protection.*

Figure 1-8. Specific explanation is provided when exceptions apply only to individual items in a list.

Informational notes are used when:

1. Informational notes provide basic information to aid the *NEC* user. Examples of this type of note include the following:
   - **90.5(C) Explanatory Material**
   - **Article 100 Definition of "Listed" Informational Note**
   - **110.11 Informational Note No. 1 and 2**
2. Informational notes provide examples of where or how the rules would apply. For example:
   - **250.20 Informational Note**
   - **250.96(B) Informational Note**
3. Informational notes provide references to other sections within the *NEC* to further explain the requirement and to aid the user in its proper application. For example:
   - **90.7 Informational Note No. 1, 2 and 3**
   - **250.30(A) Informational Note**
   - **314.15 Informational Note No. 1 and 2**

The rule stated in **Section 300.12** requires that all raceways, cable armors, and cable sheaths be continuous between boxes, fittings, or other enclosures or outlets. **Exception No. 1** allows short sections of raceways only (not cable armors and cable sheaths) to provide support and/or protection of cable assemblies. For example, this exception would permit a short section of a raceway, such as electrical metallic tubing (EMT) or rigid metal conduit (RMC), to protect and/or support a cable assembly for a short distance where physical damage could occur to the cable assembly. Note that while the *NEC* does not state a minimum length, the term "short section" implies a piece of a raceway. **Exception No. 2** allows for the floor stub up raceway and cable penetrations into equipment not to be terminated to the enclosure.

Another example of a situation where exceptions apply: The rule stated in **250.68(A)** requires that all connections of grounding electrode conductors to grounding electrodes be accessible, with two exceptions. **Exception No. 1** allows for connections encased in concrete or buried being inaccessible. **Exception No. 2** allows for exothermic or irreversible connections to structural steel to be covered in fireproofing materials.

4. Informational notes provide reference to other codes or standards to further explain the requirement, to inform the user of building code requirements, and to aid the user in its proper application. For example:
   - **110.16(B) Informational Note No. 1 and 2**
   - **210.52 Informational Note**
   - **300.21 Informational Note**
5. Informational notes provide a suggestion for adequate performance and/or proper design. For example:
   - **210.4(A) Informational Note**
   - **210.19(A)(1) Exception Informational Note No. 4**
   - **215.2(A)(1)(b) Informational Note No. 1**
   - **220.61(C)(2) Informational Note No. 1 and 2**

Following is an example of an informational note.

> **250.20 Alternating-Current Systems to Be Grounded.** Alternating-current systems shall be grounded as provided for in 250.20(A), (B), (C), or (D). Other systems shall be permitted to be grounded. If such systems are grounded, they shall comply with the applicable provisions of this article.
>
> Informational Note: An example of a system permitted to be grounded is a corner-grounded delta transformer connection. See 250.26(4) for conductor to be grounded.

Material for informational notes is included in the *NEC* only where the *Code*-Making Panel believes that the information is necessary for proper application of the rule(s). In most cases, the text of an informational note is indispensable information for the user and for proper application of the *NEC*. Always read the informational notes. For example, **Article 100 Definitions** includes the following informational note for the definition of AHJ:

> Informational Note: The phrase "authority having jurisdiction," or its acronym **AHJ**, is used in **NFPA** documents in a broad manner, since jurisdictions and approval agencies vary, as do their responsibilities. Where public safety is primary, the authority having jurisdiction may be a federal, state, local, or other regional department or individual such as a fire chief; fire marshal; chief of a fire prevention bureau, labor department, or health department; building official; electrical inspector; or others having statutory authority...

MANDATORY RULES, PERMISSIVE RULES, AND EXPLANATORY INFORMATION

The *NEC* follows specific rules that clearly illustrate whether requirements are mandatory or permissive. These rules also identify text that is explanatory in nature and provided to aid the *Code* user. **Article 90 Introduction** provides the method used to determine the applicability of all text. **Section 90.5** clearly defines the application of all rules and material in the *NEC*.:

> **90.5 Mandatory Rules, Permissive Rules, and Explanatory Material.**
>
> **(A) Mandatory Rules.** Mandatory rules of this *Code* are those that identify actions that are specifically required or prohibited and are characterized by the use of the terms shall or shall not.
>
> **(B) Permissive Rules.** Permissive rules of this *Code* are those that identify actions that are allowed but not required, are normally used to describe options or alternative methods, and are characterized by the use of the terms shall be permitted or shall not be required.
>
> **(C) Explanatory Material.** Explanatory material, such as references to other standards, references to related sections of this *Code*, or information related to a *Code* rule, is included in the *Code* in the form of informational notes. Such notes are informational only and are not enforceable as requirements of this *Code*.

Brackets containing section references to another NFPA document are for informational purposes only and are provided as a guide to indicate the source of the extracted text. These bracketed references immediately follow the extracted text.

Informational Note: The format and language used in this *Code* follows guidelines established by NFPA and published in the *NEC Style Manual*. Copies of this manual can be obtained from NFPA.

(D) Informative Annexes. Nonmandatory information relative to the use of the *NEC* is provided in informative annexes. Informative annexes are not part of the enforceable requirements of the *NEC*, but are included for information purposes only.

Mandatory Language

As stated in **90.5(A)**, a rule is considered mandatory by the use of the terms *shall* or *shall not*. Where either term is used in the *NEC*, the rule is mandatory unless an exception follows or the rule exists in Chapters 1 through 4 and is supplemented or modified in Chapter 5, 6, or 7.

230.6 Conductors Considered Outside the Building. Conductors shall be considered outside of a building or other structure under any of the following conditions:

(1) Where installed under not less than 50 mm (2 in.) of concrete beneath a building or other structure

(2) Where installed within a building or other structure in a raceway that is encased in concrete or brick not less than 50 mm (2 in.) thick

(3) Where installed in any vault that meets the construction requirements of Article 450, Part III

(4) Where installed in conduit and under not less than 450 mm (18 in.) of earth beneath a building or other structure

(5) Where installed within rigid metal conduit (Type RMC) or intermediate metal conduit (Type IMC) used to accommodate the clearance requirements in 230.24 and routed directly through an eave but not a wall or a building.

The preceding *Code* rule in **Article 230 Services**, in **Part I. General** requires that service conductors meet any of the conditions in list items 1 through 5 be considered as being outside of the building.

Permissive Text

As stated in **90.5(B)**, a *Code* rule is considered permissive by the use of the terms "shall be permitted" or "shall not be required." Where these terms are used in the *NEC*, the *Code* rule is permissive unless an exception follows the rule or unless the rule exists in Chapters 1 through 4 and is supplemented or modified in Chapters 5 through 7.

250.20 Alternating-Current Systems to Be Grounded. Alternating-current systems shall be grounded as provided for in 250.20(A), (B), (C), or (D). Other systems shall be permitted to be grounded. If such systems are grounded, they shall comply with the applicable provisions of this article.

Informational Note: An example of a system permitted to be grounded is a corner-grounded delta transformer connection. See 250.26(4) for conductor to be grounded.

Using another example, the requirement in **Article 250 Grounding** in **Part II. System Grounding** mandates the following:

1. Alternating-current systems shall be grounded as provided for in **250.20(A)**, **(B)**, **(C)**, or **(D)**, and
2. Other systems shall be permitted to be grounded. If such systems are grounded, they shall comply with the applicable provisions of this article.

The second requirement of this section specifically permits systems other than those illustrated in **250.20(A)**, **(B)**, **(C)**, or **(D)** to be grounded by the use of the term "shall be permitted." This wording is, in essence, a mandated permission. The alternative would be for the *NEC* to use the term "may" to illustrate permissiveness. However, "may" is considered unenforceable and is not used because it could be taken to mean that the inspector "may" permit other systems to be grounded or he/she "may not." The *NEC Style Manual* lists terms to be avoided, such as "may," which are vague and/or unenforceable, in an effort to make the *Code* as clear and usable as possible. **See Figure 1-9.**

Other vague terms the *NEC* avoids include: *acceptable, adequate, appropriate, desirable, familiar, frequent, generally, likely, most, near, practices, prefer, proper, reasonable, satisfactory, significant, sufficient, suitable*, and *workmanlike*.

Figure 1-9 *NEC* Text Rules

TYPE OF TEXT	LISTED TERM(S)	
Mandatory Text	"Shall" "Shall not"	
Permissive Text	"Shall be permitted"	"Shall not be required"
Explanatory Text	Informational Notes	Informative Annex

Figure 1-9. Text language in the NEC *includes mandatory, permissive, and explanatory rules*

Units of Measurement

The *NEC* uses both the SI system (metric units) and inch-pound units. The SI system value is always shown first, followed by the inch-pound value in parentheses. For example:

334.30 Securing and Supporting. Nonmetallic-sheathed cable shall be supported and secured by staples; cable ties listed and identified for securement and support; or straps, hangers, or similar fittings designed and installed so as not to damage the cable, at intervals not exceeding 1.4 m (4 ½ ft) and within 300 mm (12 in.) of every cable entry into enclosures such as outlet boxes, junction boxes, cabinets, or fittings. Flat cables shall not be stapled on edge.

Sections of cable protected from physical damage by raceway shall not be required to be secured within the raceway. Note also that **90.9(D)** specifically permits the use of either the SI system or the inch-pound system.

Extracted Material

NFPA 70: National Electrical Code is one of many documents published by the National Fire Protection Association. When another NFPA document has primary jurisdiction over material to be included in the *NEC*, the material is extracted into the *NEC*. An example of another NFPA document that would have primary jurisdiction over material addressed by the *NEC* is *NFPA 20-2010: Standard for the Installation of Stationary Pumps for Fire Protection*. When such extraction occurs, the document in which the extracted material exists is identified at the beginning of the article. An informational note immediately follows the title of the article to inform the *Code* user of the presence of extracted material. Rules within the article that are extracted are followed with the title of the referenced NFPA document and sections in brackets. Referencing sections of another NFPA document is for informational purposes only, as stated in **90.5(C)**. An example of extract material in the *NEC* can be seen in **Article 695**:

ARTICLE 695
Fire Pumps

695.3 Power Sources for Electric Motor-Driven Fire Pumps

(3) Dedicated Feeder. A dedicated feeder shall be permitted where it is derived from a service connection as described in 695.3(A)(1). [**20**:9.2.2(3)]

The informational note following the title of **Article 695** informs the user that it contains extract material from *NFPA 20-2010: Standard for the Installation of Stationary Pumps for Fire Protection*. The informational note further explains that where this material is located, it is identified with a reference in brackets at the end of the rule. **See Figure 1-10.** This extracted material can only be changed editorially to fit the style of the *NEC*.

Figure 1-10 Bracketed Text Throughout the *NEC*

ARTICLE 695
Fire Pumps

695.1 Scope.

Informational Note: Text that is followed by a reference in brackets has been extracted from NFPA 20-2013, *Standard for the Installation of Stationary Pumps for Fire Protection*. Only editorial changes were made to the extracted text to make it consistent with this *Code*.

(A) Covered. This article covers the installation of the following:

(1) Electric power sources and interconnecting circuits
(2) Switching and control equipment dedicated to fire pump drivers

(B) Not Covered. This article does not cover the following:

Reprinted with permission from NFPA 70-2017, *National Electrical Code®*, Copyright© 2016, National Fire Protection Association, Quincy, MA 02169. This reprinted material is not the complete and official position of the NFPA on the referenced subject, which is represented only by the standard in its entirety.

Figure 1-10. Throughout the NEC*, references are made to other NFPA documents that have primary jurisdiction over the material.*

ARTICLE 695
Fire Pumps
695.10 Listed Equipment. Diesel engine fire pump controllers, electric fire pump controllers, electric motors, fire pump power transfer switches, foam pump controllers, and limited service controllers shall be listed for fire pump service. [**20:** 9.5.1.1, 10.1.2.1, 12.1.3.1]

The bracketed text located at the end of **Section 695.10** identifies the sections and title of the document from which the extracted material originated.

Cross-reference Tables

The *NEC* provides six cross-reference tables where the information in a *Code* article may be specifically addressed by another article in the *Code*. In some cases, multiple modifications or supplemental requirements related to the scope of an article are located elsewhere in the *NEC*. **See Figure 1-11.**

Figure 1-11 Cross-Reference Tables

TYPE OF TEXT	LISTED TERMS
ARTICLE 210 Branch Circuits **Table 210.2 Specific-Purpose Branch Circuits**	Provides cross-references to aid the *Code* user in 9 other locations.
ARTICLE 220 Branch-Circuit Feeder and Service Calculations **Table 220.3 Additional Load Calculation Reference**	Provides cross-references to aid the *Code* user in 11 other locations.
ARTICLE 225 Outside Branch Circuit and Feeders **Table 225.3 Other Articles**	Provides cross-references to aid the *Code* user in 25 other articles.
ARTICLE 240 Overcurrent Protection **Table 240.3 Other Articles**	Provides cross-references to aid the *Code* user in 35 other articles.
ARTICLE 250 Grounding and Bonding **Table 250.3 Additional Grounding Requirements**	Provides cross-references to aid the *Code* user in over 90 other locations.
ARTICLE 430 Motors, Motor Circuits, and Controllers **Table 430.5 Other Articles**	Provides cross-references to aid the *Code* user in 26 other locations.

Figure 1-11. There are six cross-reference tables in the NEC *that list multiple changes to the scope of an article.*

FACT

The term "Tap Conductor" as defined in **Article 240.2** does not apply to service conductors used to connect fire pumps. Fire pump overcurrent protection provides short-circuit and ground fault protection to the conductors.

Table 225.3 is a cross-reference table for outside branch circuit and feeders installations for various locations. **See Figure 1-12.**

Outlines, Diagrams, and Drawings

The *NEC* does not contain pictures or commentary to aid the *Code* user in the application of installation requirements. **90.1(C)** clearly states that the *NEC* is not intended as a design manual or as an instruction manual for untrained persons. However, the *NEC* does include outlines, diagrams, and drawings that describe the application of an article or section or provide basic information in ladder-type diagrams to aid the *Code* user. **See Figure 1-13.**

90.3 Code Arrangement.
Figure 90.3 Code Arrangement.

The illustration provided with **90.3** is designed to clearly outline the arrangement and application of rules contained in the *NEC*.

210.52 Dwelling Unit Receptacle Outlets.
Figure 210.52(C)(1) Determination of Area Behind a Range, Counter-Mounted Cooking Unit or Sink.

Figure 1-12 Cross-Reference Tables

Table 220.102 Method for Calculating Farm Loads for Other Than Dwelling Unit

Ampere Load at 240 Volts Maximum	Demand Factor (%)
The greater of the following:	
All loads that are expected to operate simultaneously, or	100
125 percent of the full load current of the largest motor, or	
First 60 amperes of the load	
Next 60 amperes of all other loads	50
Remainder of other loads	25

220.103 Farm Loads — Total. Where supplied by a common service, the total load of the farm for service conductors and service equipment shall be calculated in accordance with the farm dwelling unit load and demand factors specified in Table 220.103. Where there is equipment in two or more farm equipment buildings or for loads having the same function, such loads shall be calculated in accordance with Table 220.102 and shall be permitted to be combined as a single load in Table 220.103 for calculating the total load.

Table 220.103 Method for Calculating Total Farm Load

Individual Loads Calculated in Accordance with Table 220.102	Demand Factor (%)
Largest load	100
Second largest load	75
Third largest load	65
Remaining loads	50

Note: To this total load, add the load of the farm dwelling unit calculated in accordance with Part III or IV of this article. Where the dwelling has electric heat and the farm has electric grain-drying systems, Part IV of this article shall not be used to calculate the dwelling load.

Table 225.3 Other Articles

Equipment/Conductors	Article
Branch circuits	210
Class 1, Class 2, and Class 3 remote-control, signaling, and power-limited circuits	725
Communications circuits	800
Community antenna television and radio distribution systems	820
Conductors for general wiring	310
Electrically driven or controlled irrigation machines	675
Electric signs and outline lighting	600
Feeders	215
Fire alarm systems	760
Fixed outdoor electric deicing and snow-melting equipment	426
Floating buildings	553
Grounding and bonding	250
Hazardous (classified) locations	500
Hazardous (classified) locations — specific	510
Marinas and boatyards	555
Messenger-supported wiring	396
Mobile homes, manufactured homes, and mobile home parks	550
Open wiring on insulators	398
Over 1000 volts, general	490
Overcurrent protection	240
Radio and television equipment	810
Services	230
Solar photovoltaic systems	690
Swimming pools, fountains, and similar installations	680
Use and identification of grounded conductors	200

Reprinted with permission from NFPA 70-2017, *National Electrical Code®*, Copyright© 2016, National Fire Protection Association, Quincy, MA 02169. This reprinted material is not the complete and official position of the NFPA on the referenced subject, which is represented only by the standard in its entirety.

Figure 1-12. Cross-reference tables refer to other articles that are applicable to the subject covered by the NEC.

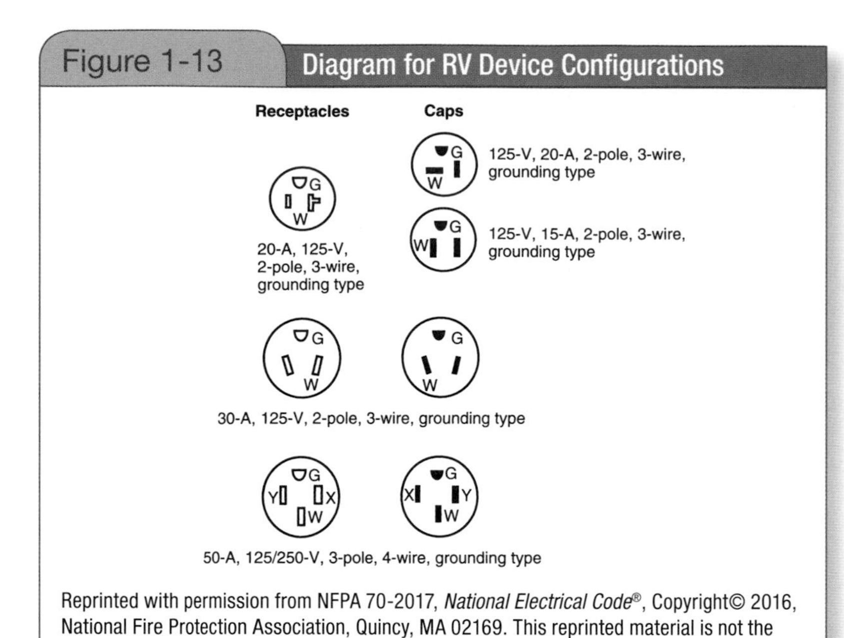

Figure 1-13. The NEC *includes many diagrams to aid the* Code *user.*

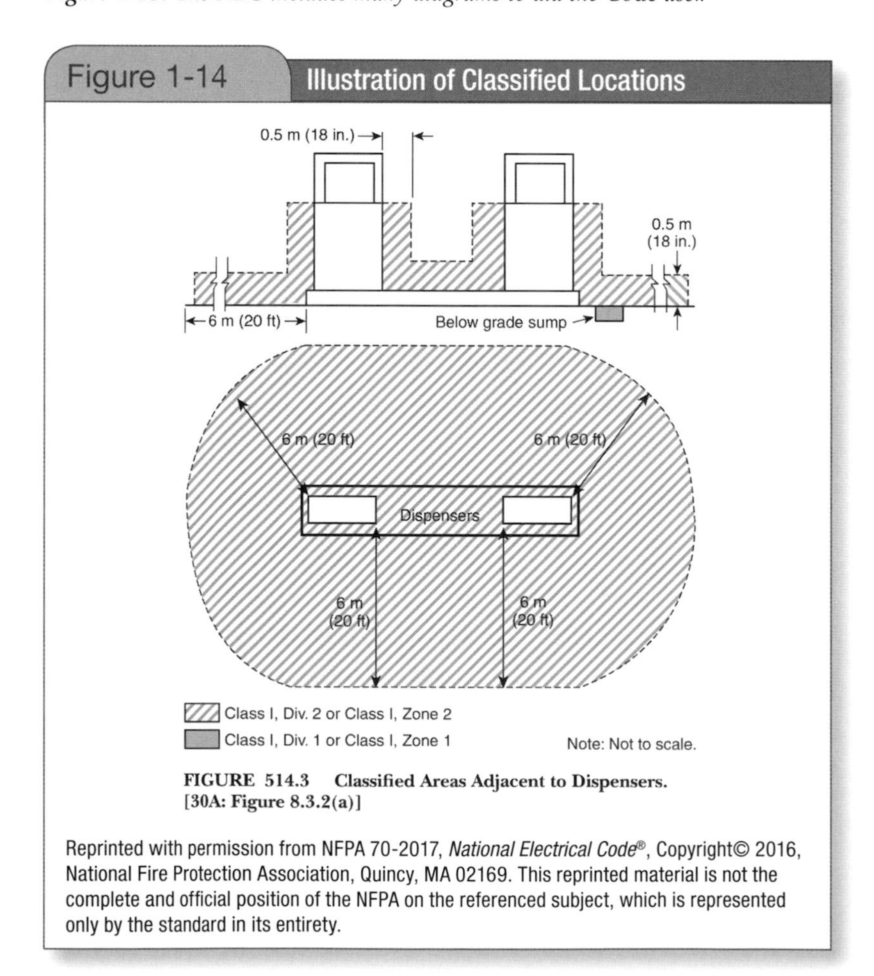

Figure 1-14. Requirements for Class I fueling facility environments are covered in Article 514.

This drawing is provided to aid the *Code* user in applying the receptacle outlet requirement of **210.52(C)(1)** near a sink, counter-mounted cooking unit, or range.

250.1 Scope.
Figure 250.1 Grounding and Bonding.

This illustration details the organization of **Article 250 Grounding and Bonding**

514.3 Classification of Locations.
Figure 514.3 Classified Areas Adjacent to Dispensers as Detailed in Table 514.3(B)(1). [30A: Figure 8.3.1]

This drawing is included to aid the *Code* user in applying **514.3(B)(1)** in classified locations adjacent to dispensers. **See Figure 1-14.** Many other examples of drawings and diagrams are used though out the *Code* to help the user identify important *Code* requirements.

DEFINITIONS AND LANGUAGE OF THE *NEC*

Article 100 contains only those definitions that are essential to the application of this *Code*. The intent is not to include commonly-defined general terms or commonly-defined technical terms from related codes and standards. In general, only those terms that are used in two or more articles are defined in **Article 100**. Other definitions are included in the article in which they are used, but may be referenced in **Article 100**. Notice in Article 100 after each definition is the number of the Code Making Panel that has the primary responsibility for the oversight of that definition.

Part I. General of **Article 100** contains definitions intended to apply wherever the terms are used throughout the *Code*. **Part II. Over 1000 Volts, Nominal** contains definitions applicable only to the parts of articles specifically covering installations and equipment operating at more than 1000 volts, nominal.

NEC Language: Overview

Many of the words used in the *NEC* have specific meaning to the article or topic addressed. Users of the *NEC* must learn the definitions of terms used within the document. The *NEC* might seem to be written in a different language; therefore, **Article 100 Definitions** will provide a framework of the terms with particular electrical connotations. Terms that are defined must follow the *NEC Style Manual*. The *Code* does not attempt to define commonly-used general terms, such as *area* or *space*, or commonly-used technical terms from other related codes and standards. Only those terms that are essential to the proper application of the *NEC* are defined. Unless a term is used in two or more articles, it will not appear in **Article 100** and will instead be defined in the article in which it is used. The rules as presented in the *NEC Style Manual* for definitions are as follows:

- Definitions are in alphabetical order.
- Definitions do not contain the term being defined.
- Defined terms that appear in two or more articles are listed in **Article 100**.
- Defined terms that appear in a single article are listed in the second section (for example, **680.2**) of that article.

All users of the *NEC* must be familiar with definitions in order to properly interpret and apply each requirement. Moreover, the *NEC* definition for what seems to be a "common term" might differ from a standard dictionary explanation, therefore seriously affecting the implementation of an electrical installation. For example:

Q. *A metal-sheathed cable – Type MC is installed above a lay-in type acoustical ceiling with the ceiling tiles in place. The MC cable is hidden behind the lay-in tiles and can no longer be seen. Is the MC cable considered to be exposed?*

A. *Yes. The MC cable above the lay-in type acoustical ceiling is installed in accordance with the NEC definition of exposed. In a standard dictionary, the term "exposed" would normally mean the item could be seen. However, the Code defines the term "exposed" in two different scenarios.*

- *Exposed (as applied to live parts)*
- *Exposed (as applied to wiring methods)*

Because this question is about type MC cable, which is a wiring method, the second definition of "exposed" would apply:

> **Article 100 Definitions.**
> **Exposed (as applied to wiring methods).** On or attached to the surface or behind panels designed to allow access. (CMP-1)

Note that this definition would apply to the type of MC cable installed above a lay-in type acoustical ceiling because it is installed behind panels (lay-in ceiling tile) designed to allow access. **See Figure 1-15.**

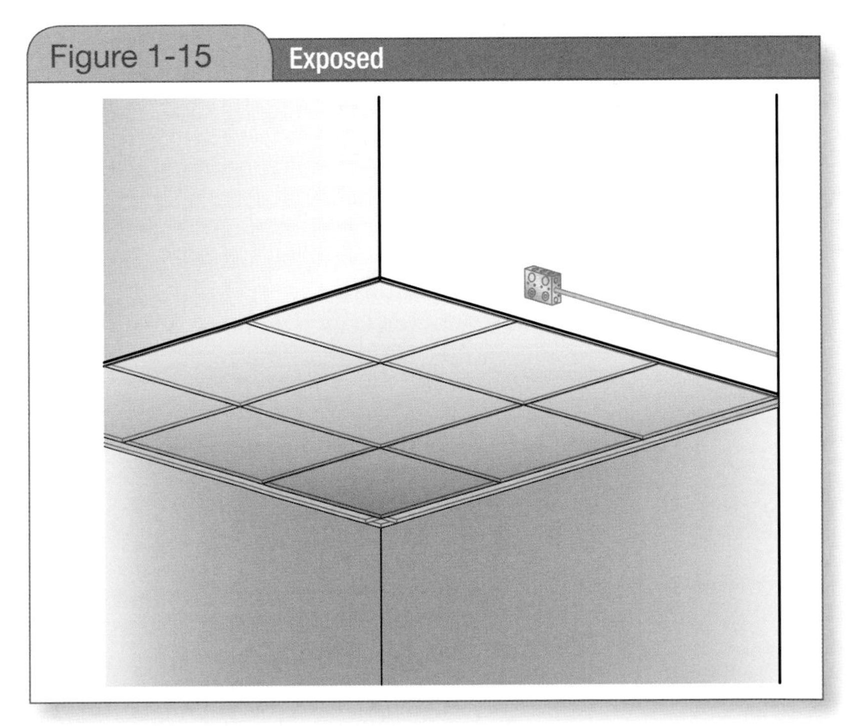

*Figure 1-15. A box above a lay-in ceiling is defined as "exposed" (as applied to wiring methods) under **Article 100 Definitions**.*

For additional Information, visit qr.njatcdb.org Item #2404

Figure 1-16 **Exposed**

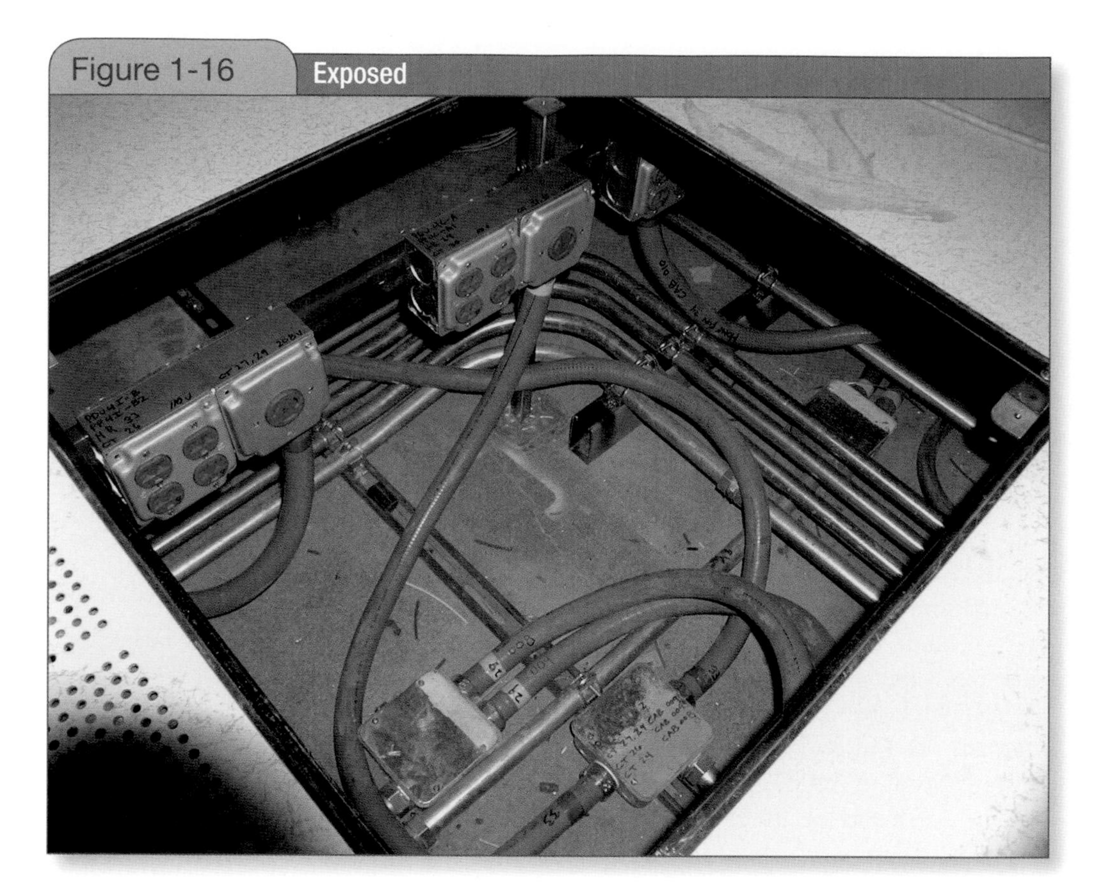

Figure 1-16. Receptacles and junction boxes below a raised computer floor are defined as "exposed" (as applied to wiring methods) under ***Article 100 Definitions.***

This definition would also apply to receptacles and junction boxes under a raised computer floor. **See Figure 1-16.** The term "exposed" is used more than 300 times in the *NEC*. All users of the *Code* must be familiar with the most common terms used throughout the document.

Many articles define terms used only within the same article. The *NEC Style Manual* requires that such terms be listed within the second section of that article, so that the *NEC* user can see early on the terms defined for proper application of the requirements that follow. The first section is always the scope of an article, and the second is a list of definitions of terms used within that article.

For example, in **Article 517 Health Care Facilities,** the second section of the article, **Section 517.2 Definitions**, lists 47 terms and definitions. Remember, a term must be defined in **Article 100** if it is used in two or more articles. **Article 100** contains more than 150 definitions.

NEC Language: Definitions

The following **Article 100** definitions are examples of the commonly used terms in the *NEC* that the *Code* user must clearly understand for proper application of its requirements.

Accessible, Readily (Readily Accessible). Capable of being reached quickly for operation, renewal, or inspections without requiring those to whom ready access is requisite to take actions such as to use tools (other than keys), to climb over or under, to remove obstacles or to resort to portable ladders, and so forth. (CMP-1)

Figure 1-17 Accessible

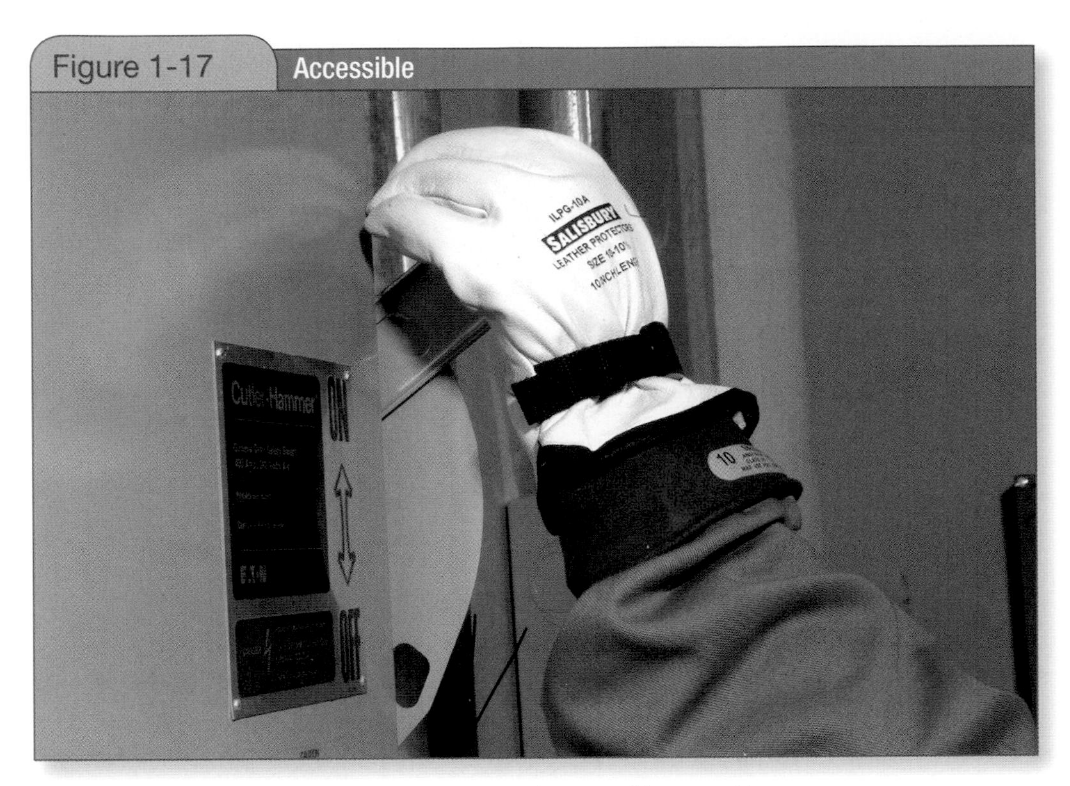

Figure 1-17. A person can quickly reach equipment that is readily accessible.

The term "readily accessible" is used more than 70 times in the *NEC*. Other terms related to accessibility and defined in **Article 100** are "accessible" (as applied to equipment), "accessible" (as applied to wiring methods), and "concealed." **See Figure 1-17.**

If a disconnect switch was located behind a stack of boxes, it would be considered accessible, not readily accessible. Also, if the disconnect switch had a padlock installed, again, it would only be considered accessible.

Ampacity. The maximum current, in amperes, that a conductor can carry continuously under the conditions of use without exceeding its temperature rating. (CMP-6)

The term "ampacity" is used more than 400 times in the *NEC* and is derived from the combination of the terms "ampere" and "capacity." **See Figure 1-18.**

Approved. Acceptable to the authority having jurisdiction. (CMP-1)

"Approved" is used more than 300 times in the *NEC*. It is essential that the *Code* user know that when this term is used, compliancy is determined by the "authority having jurisdiction" (AHJ) and not a third-party listing organization such as UL.

The term "authority having jurisdiction" is also defined in **Article 100**. The informational note following the definition states that the AHJ may be federal, state, local, or other regional departments or individuals such as a fire chief, fire marshal, labor department, health department, electrical inspector, or others having statutory authority. The note further states that for insurance purposes, an insurance inspection department, rating bureau, or other insurance company representatives may be the AHJ.

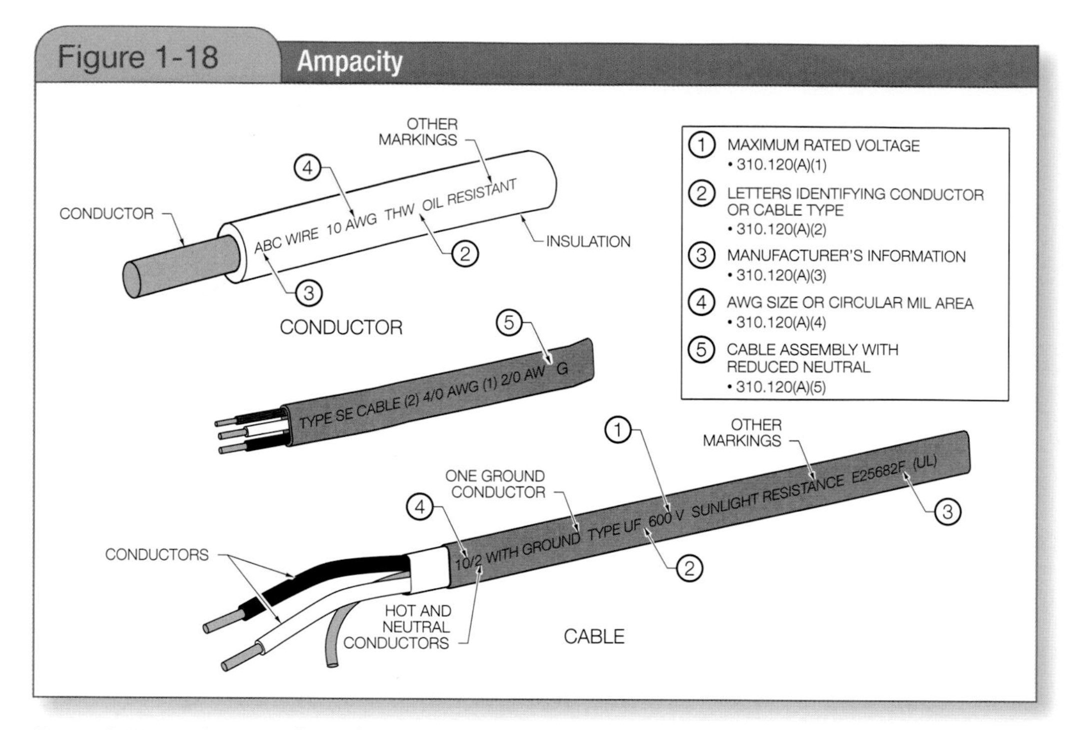

Figure 1-18. *Conductor markings do not list the actual ampacity of the conductor.*

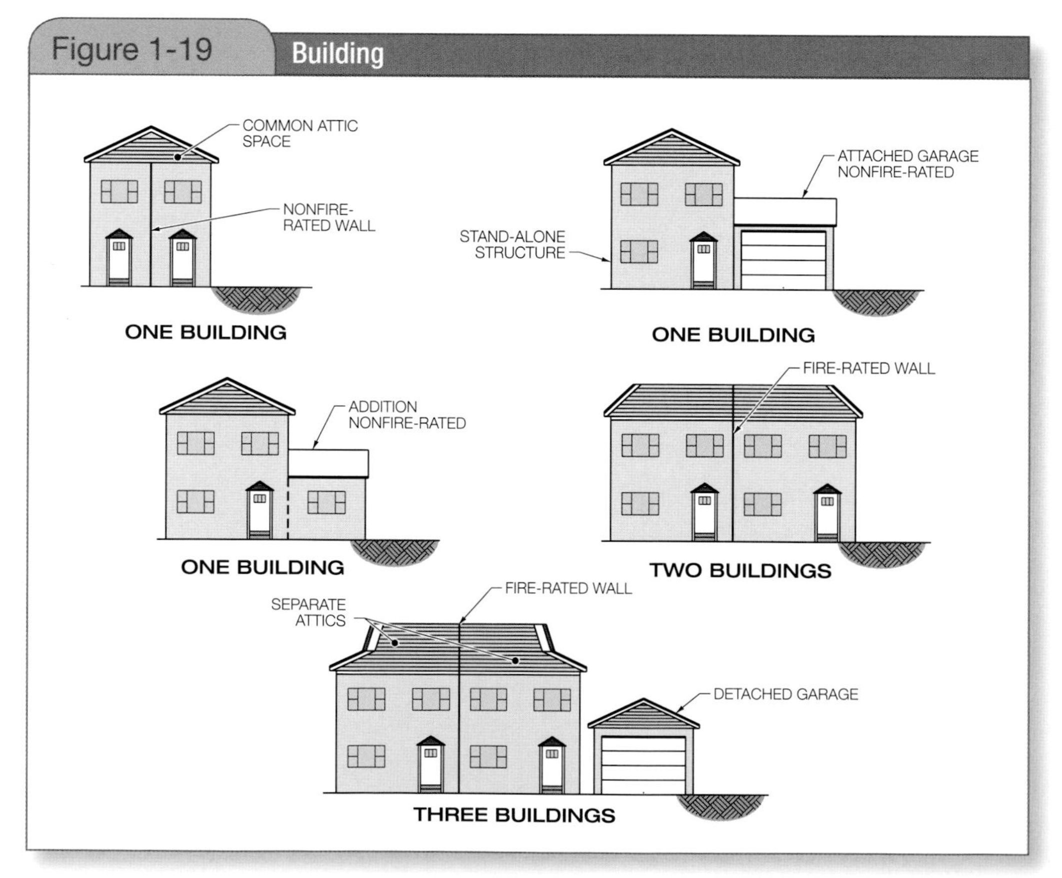

Figure 1-19. *Buildings are defined in the* NEC *as stand-alone structures or portions of a building separated by fire-rated walls.*

Branch Circuit. The circuit conductors between the final overcurrent device protecting the circuit and the outlet(s). (CMP-2)

Building. A structure that stands alone or that is cut off from adjoining structures by fire walls with all openings therein protected by approved fire doors. (CMP-1)

The *NEC* contains specific requirements and limitations for buildings, so it is essential to understand that a row of ten strip stores, for example, may be recognized in the *NEC* as 10 separate buildings if they are separated by fire walls. **See Figure 1-19.**

Device. A unit of an electrical system, other than a conductor, that carries or controls electric energy as its principal function. (CMP-1)

The term "device" is used in the *NEC* more than 500 times. Types of devices would include but not be limited to, receptacles and switches. **See Figure 1-20.**

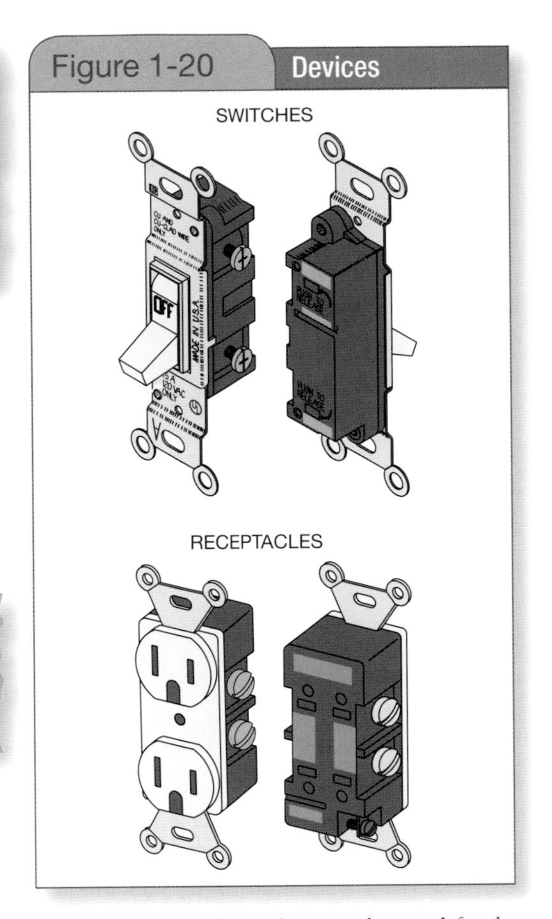

Figure 1-20. *Switches and receptacles are defined as devices in* ***Article 100****.*

Feeder. All circuit conductors between the service equipment, the source of a separately derived system, or other power supply source and the final branch-circuit overcurrent device. (CMP-2)

For proper application of the *NEC*, the *Code* user must be able to determine the proper *Code* term for all current-carrying conductors. The four types of current-carrying conductors are branch circuit, feeder, service, and tap conductors.

Fitting. An accessory such as a locknut, bushing, or other part of a wiring system that is intended primarily to perform a mechanical rather than an electrical function. (CMP-1)

Fittings are used to terminate conduit and other raceways to boxes and enclosures. **See Figure 1-21.**

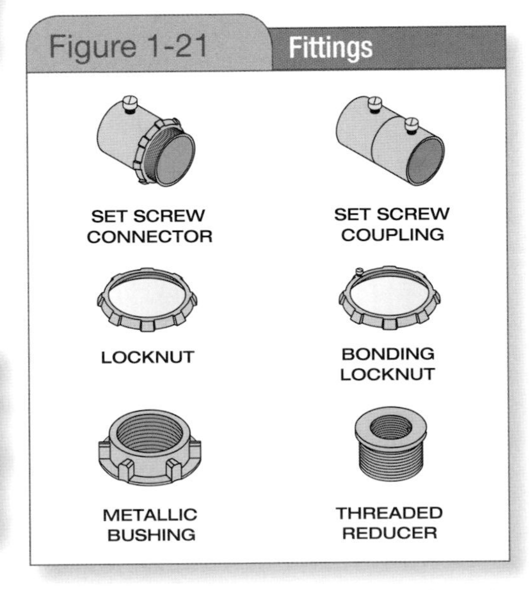

Figure 1-21. *Fittings only perform a mechanical function, such as termination of conduit to an enclosure.*

In Sight From (Within Sight From, Within Sight). Where this Code specifies that one equipment shall be "in sight from," "within sight from," or "within sight of," and so forth, another equipment, the specified equipment is to be visible and not more than 15 m (50 ft) distant from the other. (CMP-1)

When the *NEC* requires that equipment such as a motor and an associated disconnecting means be "within sight of" each other, the requirement is that the equipment be visible and not more than 50 feet apart.

Outlet. A point on the wiring system at which current is taken to supply utilization equipment. (CMP-1)

The term "outlet" would include a receptacle, a ceiling-mounted box for a lighting fixture, and the point at which equipment is hard-wired. **See Figure 1-22.**

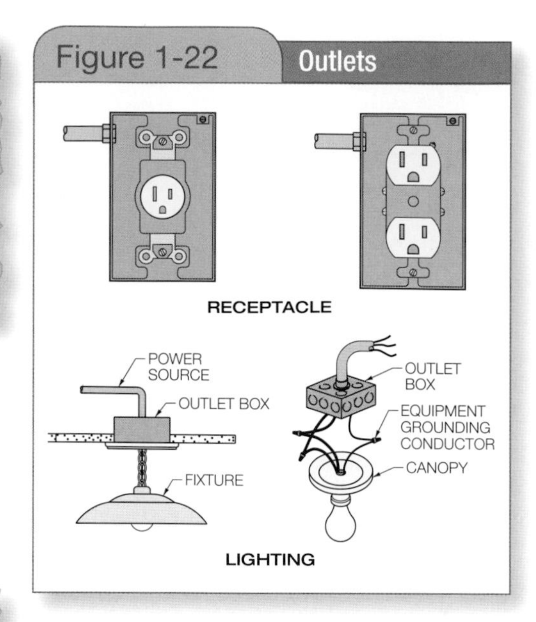

Figure 1-22. *Outlets are not only receptacles, but include lighting outlets in the ceiling.*

Overcurrent. Any current in excess of the rated current of equipment or the ampacity of a conductor. It may result from overload, short circuit, or ground fault. (CMP-10)

An overcurrent includes the following:

- Overload: above the normal full-load rating, the current stays in the normal current path. For example, 22 amperes flowing on a 20-ampere branch circuit is an overload. Too many appliances or equipment may be plugged into the circuit.
- Short Circuit: (Not defined in **Article 100** or elsewhere in the *Code*) when the current does not flow through its normal path (takes a short cut) but continues on the circuit conductors with contact from line-to-line (black ungrounded conductor touching a red ungrounded conductor of a different circuit) or line-neutral contact.

Qualified Person. One who has skills and knowledge related to the construction and operation of the electrical equipment and installations and has received safety training to recognize and avoid the hazards involved. (CMP-1)

The term "qualified person" is used more than 100 times in the *NEC*. It is extremely important that all *Code* users understand that a qualified person has specific skills and knowledge as well as safety training.

The informational note refers the user to *NFPA 70E: Standard for Electrical Safety in the Workplace* for electrical safety training requirements.

Summary

The *Codeology* process is based on the four basic building blocks found in the *NEC*:

1. The Contents pages of the *NEC*
2. Section 90.3 Code Arrangement
3. The structure of the *NEC*
4. Definitions—the language of the *NEC*

Marking article titles, parts, and selected section information will allow the *Code* user to quickly find information in the book. One of the most basic and crucial steps to becoming a proficient user of the *NEC* is to gain an in-depth understanding of the structure of the *NEC* text.

The *NEC Style Manual* governs the structure of the text through a detailed outline form followed consistently throughout the *Code*. An understanding of this outline form is essential to proper application of *NEC* requirements. The *NEC* is divided into 10 major parts: the introduction and nine chapters. The *Code* is organized into chapters, parts, articles, sections, subsections, exceptions, and informational notes common to installation requirements. Parts are major subdivisions of articles that logically separate information for ease of use and proper application.

Parts are then broken down into separate sections and individually titled to address the scope of the individual part. Sections may be logically subdivided into three levels. Sections may also contain list items, exceptions, and informational notes. Understanding the language and definitions used in the *Code* is vital to properly interpreting the sections.

The *Code* is arranged so Chapters 1 through 4 apply generally to all installations and Chapters 5 through 7 address "special" installations that can be more restrictive than found in Chapters 1 through 4. Chapter 8, Communications, does not apply to Chapters 1 to 7 unless Chapter 8 specifically references an article or section in Chapters 1 to 7, and Chapter 9 provides tables referenced in other articles. Using the *Codeology* method will allow for quick access to information necessary for success in the electrical industry.

The structure of the *NEC* is in a progressive, ladder-type format, which when applied is as follows:

- The rule in the second-level subdivision, which is limited to:
- The rule in the first-level subdivision, which applies only under:
- The section in which it exists, which applies only in:
- The part it is arranged in, which applies only in:
- The article in which it exists, which is limited to:
- The chapter in which it is located.
- List items are used in sections, subdivisions, or exceptions where necessary.
- Exceptions are used only when absolutely necessary and are always italicized.
- Informational notes are informational only and are not mandatory.
- Tables in Chapter 9 are applicable only where referenced elsewhere in the *NEC*.
- Informative Annexes are informational only and are not mandatory.

Summary

Mandatory language in the *NEC* consists of the use of “shall” and “shall not.” Permissive language in the *NEC* consists of the use of “shall be permitted” and “shall not be required.” Explanatory text exists in the form of informational notes. Cross-references, outlines, drawings, and diagrams are included to aid the user in the proper application of the *NEC*.

An in-depth understanding of the structure or outline form of the *NEC* is necessary to properly apply the requirements of the *Code*. Understanding this structure or outline form of the *NEC* is one of the cornerstones of the *Codeology* method.

Review Questions

1. **Where multiple modifications or supplemental requirements are related to the scope of an article located elsewhere in the *NEC*, an article will contain a __?__ table to aid the *Code* user.**
   a. calculations
   b. contents
   c. conduit fill
   d. cross-reference

2. **Which of the following is an example of permissive language in the *NEC*?**
   a. "If the installer desires"
   b. "May be permitted"
   c. "May if cost is an issue"
   d. "Shall be permitted"

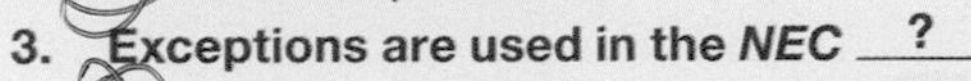

3. **Exceptions are used in the *NEC* __?__.**
   a. only when necessary
   b. to allow alternative methods
   c. to confuse the *Code* user
   d. to justify shortcuts

4. **Informational notes are __?__.**
   a. designed to aid the *Code* user
   b. explanatory material
   c. informational only
   d. all of the above

5. **When an article is subdivided into logical separations, these subdivisions are called __?__.**
   a. annexes
   b. parts
   c. sections
   d. subdivisions

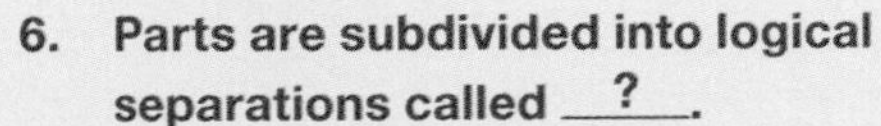

6. **Parts are subdivided into logical separations called __?__.**
   a. informative annexes
   b. parts
   c. sections
   d. subdivisions

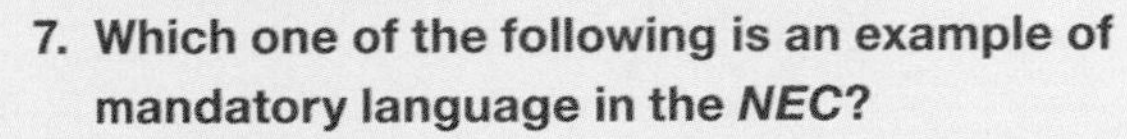

7. **Which one of the following is an example of mandatory language in the *NEC*?**
   a. "May not"
   b. "Must not"
   c. "Shall not"
   d. "Should not"

Review Questions

8. Which of the following types of informational notes are designed to aid the user of the *NEC*?

a. Design suggestions and/or examples
b. Informational
c. Reference
d. All of the above

9. The table of contents is broken down into __?__ major subdivisions.

a. 2
b. 8
c. 9
d. 10

10. The Introduction to the *NEC* is located in __?__.

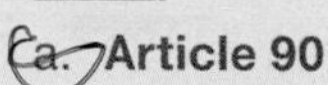
a. Article 90
b. Chapter 1
c. the preface
d. the table of contents

11. Articles within the *NEC* in the 500 series are dedicated in scope to __?__.

a. special conditions
b. special occupancies
c. utilization equipment
d. wiring methods

12. Which chapter will contain an Article to address the special equipment requirements for swimming pools?

a. 5
b. 6
c. 7
d. 8

13. The answer to a question about the size of conductors supplying a motor in a general application will be found in Chapter __?__.

a. 1
b. 2
c. 3
d. 4

Review Questions

14. The tables located in Chapter 9 of the *NEC* apply __?__.

a. as referenced in the *NEC*

b. at all times only in Chapters 5, 6, and 7

c. only in Chapters 5, 6, and 7

d. wherever they are useful

15. Which chapter of the *NEC* stands alone and is not subject to any other of its chapters unless a specific reference is made?

a. Chapter 1

b. Chapter 5

c. Chapter 8

d. Chapter 9

16. Chapters 1 through 4 of the *NEC* apply __?__ to all electrical installations.

a. generally

b. in some cases

c. sparingly

d. without modification

17. Annexes in Chapter 9 of the *NEC* are __?__.

a. applicable as referenced

b. informational only

c. mandatory requirements

d. used only for special equipment

18. Chapters 5, 6, and 7 of the *NEC* __?__ Chapters 1 through 7.

a. apply generally with

b. are not be associated in any way with

c. do not have an effect on

d. supplement or modify

19. The arrangement of the *NEC* is outlined in Section __?__ of the *Code*.

a. 90.1

b. 90.3

c. 110.3(B)

d. 210.8

The *Codeology* Method: Fundamentals and the "General" Chapter of the *NEC*

Introduction

The *Codeology* title for Chapter 1 is "General." All electrical installations are subject to these general rules. For example, the definitions given in **Article 100** apply to the terms wherever they are used in the *Code*. General requirements also exist in the "General" chapter in **Article 110**, covering basic needs that are common to all electrical installations. In accordance with the *Code* arrangement requirements of **Section 90.3**, Chapter 1 applies to all electrical installations unless supplemented or modified by Chapters 5, 6, or 7.

Objectives

» Describe the *Codeology* method.

» Identify clues and key words to locate the proper chapter, article, and part within the *NEC*.

» Effectively use the index, key words, and clues to find information in the *NEC*.

» Name the four basic building blocks of *Codeology*.

» Recognize the *Codeology* outline in the *NEC*.

» Implement the fundamentals of *Codeology*.

» Apply the *Codeology* method to Chapter 1 General, in the *NEC*.

» Label the *Codeology* title for *NEC* Chapter 1 as "General."

» Summarize the general type of information and requirements contained in Chapter 1.

» Identify Chapter 1 numbering as the 100-series.

» Recognize, recall, and apply the articles contained in Chapter 1.

Chapter 2

Table of Contents

WHAT IS THE *CODEOLOGY* METHOD?

One of the more challenging aspects of learning the electrical trade is mastering the *National Electrical Code.* **Article 90.1(A)** informs the user that the *NEC* is not a design manual or a book used as an instruction manual for untrained persons. Using the *Code* can be a difficult and frustrating task for those who do not understand its layout and content. Jobsite productivity and profitability can be severely impacted when an inexperienced designer or worker unknowingly violates an *NEC* rule. Using the *Codeology* method can help designers, contractors, foremen, workers, and even candidates for state qualifying exams use the *Code* proficiently, finding applicable *Code* references and answers to questions quickly.

Electrical literacy is very important to the *Code* user. Those entering the electrical trade can expect a lifetime of learning, as innovative new technologies, electrical equipment, and devices come to the market every year. The initial process of learning the *NEC* and how to find information in the book is not easy; it takes considerable time to learn both its content and application. Because the *NEC* is updated every three years to stay current with the electrical industry, users must periodically learn new information and rules through continuing education and other sources. As experience is gained in the electrical industry, the *NEC* becomes easier to use. Putting in the time now to learn the basics of *Codeology* will serve any worker well over the long run of his or her career.

The *Codeology* method will explore the use of the index, clues, and key words; the four building blocks of *Codeology*, the concepts of "plan," "build," and "use;" and how to use them to find applicable *Code* sections. Many examples to practice will be provided. Exercise patience and don't give up. Look up every reference necessary to answer the provided questions.

USING THE INDEX

The index located in the back of the *Code* book provides a key link to finding words, phrases, or topics in *Code* articles and sections relating to a topic. The key to using the index is picking a key word or topic and looking it up in the index.

Q. *Generally, are junction boxes required to be accessible?*

A. *Yes, 314.29*

The key terms used in this section are "junction boxes" and "accessible." Look up the term "junction boxes" in the index. The subtopic, accessible, is under the *NEC* section reference of **314.29**. Proceed to the referenced section to find an answer to the question.

> **314.29 Boxes, Conduit Bodies, and Handhole Enclosures to Be Accessible.** Boxes, conduit bodies, and handhole enclosures shall be installed so that the wiring contained in them can be rendered accessible without removing any part of the building or structure or, in underground circuits, without excavating sidewalks, paving, earth, or other substance that is to be used to establish the finished grade.
>
> *Exception:* Listed boxes and handhole enclosures shall be permitted where covered by gravel, light aggregate, or noncohesive granulated soil if their location is effectively identified and accessible for excavation.

Using the index method will provide an opportunity to find information, but

Practice Skill: Using the Index

Q: Hydromassage bathtubs and their associated equipment shall be located on an individual branch circuit and connected to a(n) __?__ *ground-fault circuit interrupter.*

Step 1. Select the key word(s) or phrase(s): *hydromassage bathtub* and *ground-fault circuit interrupter.*

Step 2. Go to the index and look up *ground-fault circuit interrupter.*

Step 3. Select the appropriate word or phrase from the index and go to the *NEC* section reference: bathtubs, hydromassage **680.71**.

Step 4. Go to the appropriate page and section in the *Code* book and answer the question:

A: Hydromassage bathtubs and their associated equipment shall be located on an individual branch circuit and connected to a **readily accessible** *ground-fault circuit interrupter.*

this is not always the best method to locate sections or information. Selecting key words is difficult, and not all key words lead to the correct section reference in the index. The *Codeology* method will provide a faster technique for finding proper references.

Clues to finding a proper *Code* section come in many forms. Each step of the *Codeology* method exposes clues for individual chapters, articles, and parts. Using clues or key words is essential when qualifying a question or when seeking the correct part of the correct article in the correct chapter. Clues and key words should first lead to the Table of Contents and are found throughout the following:

- Questions
- Chapter titles
- Article titles
- Titles of parts
- Titles of sections and subdivisions

See Figure 2-1.

Figure 2-1 Clues and Key Words

Basic/General	Chapter 1
Plan	Chapter 2
Build	Chapter 3
Use	Chapter 4
Specials	Chapters 5 - 7
Communications	Chapter 8
Occupancy Types	Chapters 1 - 7
Indoor	**Outdoor**
Dry Location	**Damp/Wet**
Permanent	**Temporary**
1000 Volts or Less	**Over 1000 Volts**
Ungrounded	**Grounded**
Shall	**Shall Not**

Figure 2-1. Clues and key words include terms and phrases that point the user to a specific part of the Code book.

FOUR BASIC BUILDING BLOCKS OF THE *CODEOLOGY* METHOD

Four basic building blocks form the basis of our *Codeology* method and can be summarized as outlined below:

Building Block #1: Contents Pages

- Outline of the *NEC*
- Ten major subdivisions of the *NEC*
- Introduction, **Article 90**
- Chapters 1 through 9
- Subdivision of chapters into articles within the scope of each chapter

Building Block #2: Section 90.3 *Code* Arrangement

- Chapters 1 through 4 apply generally to all installations
- Chapters 5, 6, and 7 supplement or modify Chapters 1 through 7
- Chapter 8 stands alone unless a specific reference exists
- Chapter 9 tables apply as referenced; Annexes are informational only

Building Block #3: Structure of the *NEC*

- Ten major subdivisions of the *NEC*
- Introduction, **Article 90**
- Chapters 1 through 9
- Broad area of coverage in each chapter
- Subdivision of each chapter into articles to address chapter scope
- Logical subdivision of articles into parts
- Subdivision of parts into sections
- Sections can contain three levels of subdivision
- Sections and subdivisions can contain exceptions
- Sections and subdivisions can contain list items
- Sections and subdivisions can contain informational notes
- Mandatory language: *shall* or *shall not*
- Permissive language: *shall be permitted* or *shall not be required*
- Informational material contained in informational notes and Annexes

Building Block #4: Definitions, the Language of the *NEC*

- Terms used in more than one article are defined in **Article 100**
- Terms used in a single article are defined in the second section of the article

GETTING TO THE CORRECT CHAPTER, ARTICLE, AND PART

By design, the *Codeology* method allows users to determine exactly where to begin looking in the *NEC* for the section that addresses their needs. The method begins by applying the *Codeology* outline to get to the correct chapter. Through the use of key words and clues, users can then identify the correct article and part in which to begin their inquiry into the *NEC*. These are the fundamentals of *Codeology*.

The *Codeology* method is designed to teach a systematic, disciplined approach to quickly finding information in the *NEC* by understanding and applying the outline form of the *NEC*. The use of generic terms to help the *Codeology* find the correct chapter in the contents pages is essential. The generic terms include **Basic/General, Plan, Build, Use, Specials,** and **Communications**.

The generic terms are used to guide the *Codeology* user in the correct direction to find needed information. The primary focus is on *NEC* Chapters 1 through 8. **Article 90 Introduction** lays the ground rules for the use of the *NEC*, while the tables in Chapter 9 are to be used only where referenced elsewhere in the *Code* book.

Basic/General

When a question or need for information in the *NEC* is basic or general in nature to all electrical installations, think "General" and start in **Chapter 1 General**. Chapter 1 contains two articles: (1) Definitions and (2) Requirements for Electrical Installations. **See Figure 2-2.** Key words

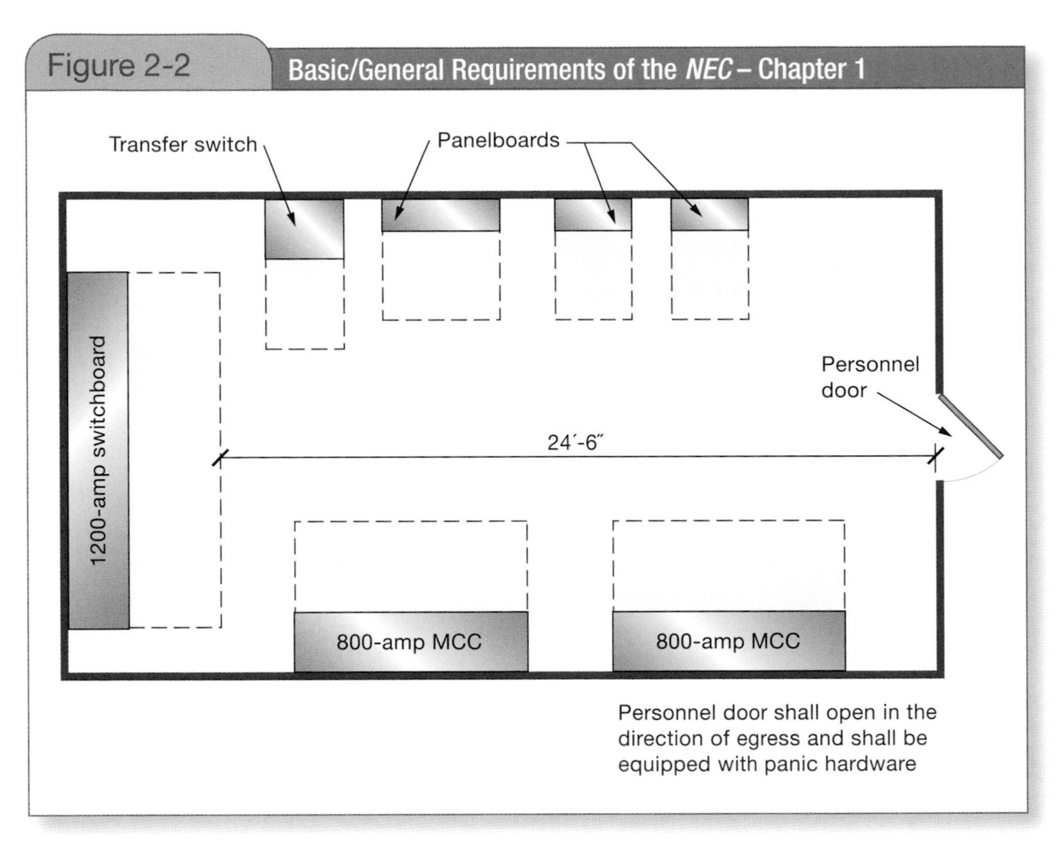

Figure 2-2. *The general requirements of Chapter 1 include provisions for adequate working space for persons and dedicated space for electrical equipment.*

and clues that should guide the *Codeology* user to think "General, Chapter 1" include the following:

- A definition question
- Examination, installation, and use of equipment (listed, labeled)
- Mounting and cooling of equipment
- Electrical connections
- Flash protection
- Identification of disconnecting means
- Workspace clearance

Plan

When a question or need for information in the *NEC* deals with planning stages and is general in nature to all electrical installations, think "Plan" and start in **Chapter 2 Wiring and Protection. See Figure 2-3.** The title of Chapter 2 includes two key words: "wiring" and "protection."

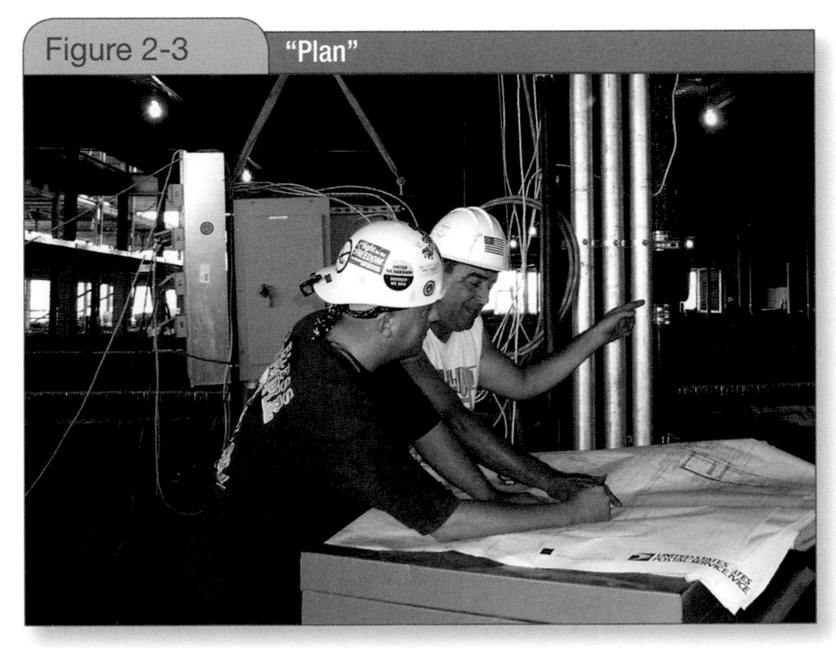

Figure 2-3. *Chapter 2 of the* NEC *is called the "Plan" chapter. All electrical installations must be properly planned.*

The term "wiring," as used in Chapter 2, does not imply different types of cable assemblies or raceways. Wiring in Chapter 2 addresses the *NEC* terms for the types of current-carrying conductors common to all electrical installations. These wiring types include branch circuits, feeders, services, taps, and transformer secondary conductors. Chapter 2 provides the basic requirements for all of these conductors and the calculations to properly size them.

The term "protection," as used in Chapter 2, includes the basic protection requirements for all electrical installations, including overcurrent protection, grounding, bonding, surge arresters, and surge-protective devices.

Key words and clues that should guide the *Codeology* user to think "Plan, Chapter 2" include the following:

- Use and identification of grounded conductors
- Branch circuit
- Feeder
- Service
- Calculation or computed load
- Overcurrent protection
- Grounding
- Surge arresters
- Surge-protective devices (SPDs), 1000 Volts or less

Build

When a question or need for information in the *NEC* deals with building an electrical installation, think "Build" and start in Chapter 3. All inquiries regarding methods and materials to get from the source of energy to the load (all physical wiring of an installation) require wiring methods and/or materials covered in **Chapter 3 Wiring Methods and Materials**. The title of Chapter 3 includes two terms: "wiring methods" and "wiring materials," which cover all means and methods of electrical distribution. Chapter 3 does not cover panelboards, disconnects, transformers, or utilization equipment. However, it does cover every physical means of wiring branch circuits, feeders, services, taps, and transformer secondary conductors.

The wiring methods portion of Chapter 3 includes general information for all wiring methods and materials as well as specific articles for all conductors, cable assemblies, raceways, busways, cablebus, and the like. In addition, the wiring materials portion of Chapter 3 includes general information for all wiring materials as well as specific articles for all boxes (outlet, device, pull, junction), conduit bodies, cabinets, cutout boxes, auxiliary gutters, and wireways. Key terms and clues that should guide the *Codeology* user to think "Build, Chapter 3" include the following:

- General questions on wiring methods or materials
- General questions for conductors, uses permitted, ampacity, etc.
- Installation, uses permitted, or construction of any cable assembly
- Installation, uses permitted, or construction of any raceway
- Installation, uses permitted, or construction of any other distribution method
- Requirements for boxes of all types
- Requirements for cabinets, meter sockets, wireways, etc.
- Open wiring on insulators and outdoor overhead conductors over 1000 volts

See Figure 2-4.

Use

When a question or need for information in the *NEC* deals in general with electrical equipment that uses, controls, or transforms electrical energy, think "Use" and start in Chapter 4. All inquiries for information on electrical equipment that controls, transforms, utilizes, or aids in the utilization of electrical energy can be found in **Chapter 4 Equipment for General Use**.

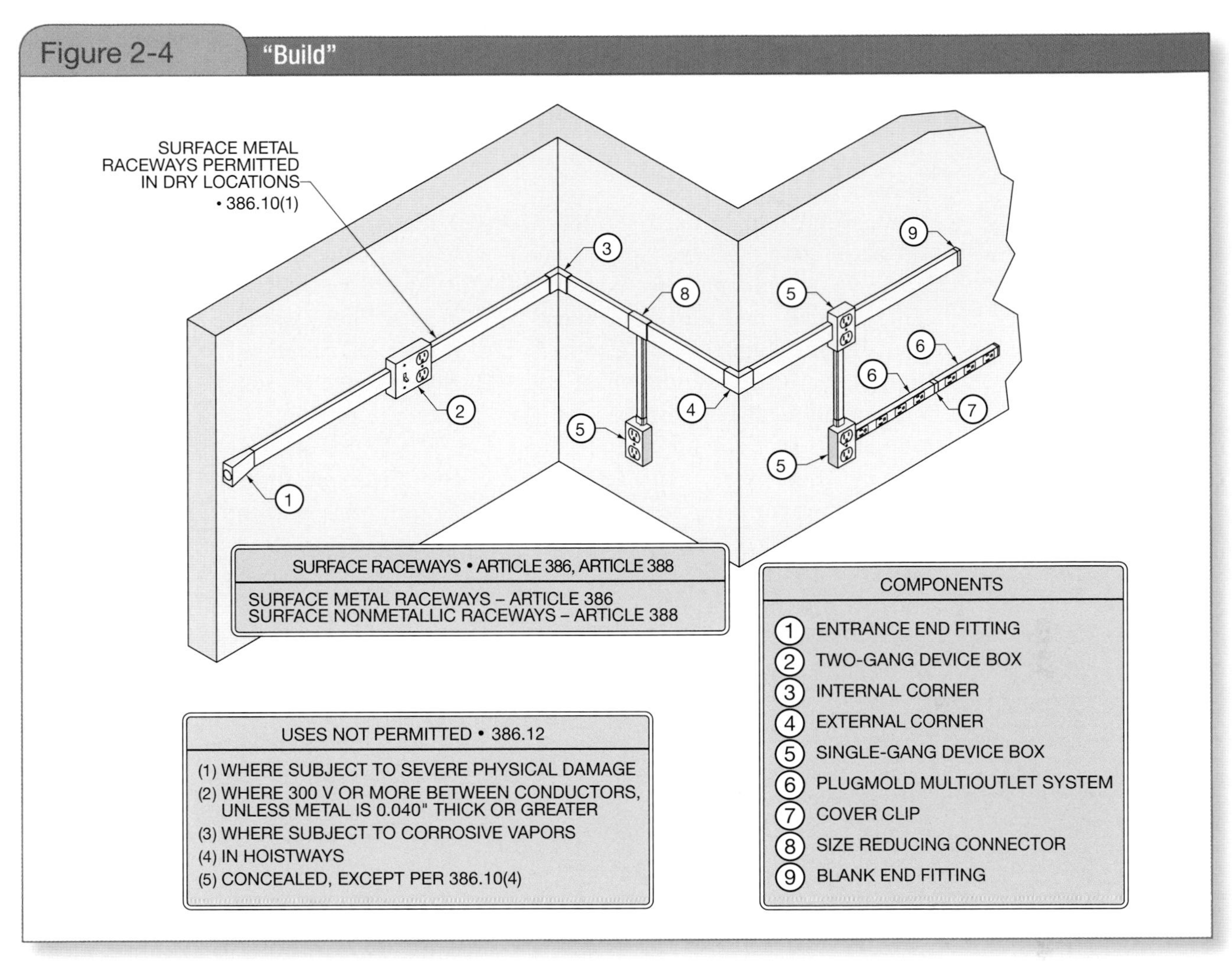

Figure 2-4
"Build"
SURFACE METAL RACEWAYS PERMITTED IN DRY LOCATIONS
• 386.10(1)
SURFACE RACEWAYS • ARTICLE 386, ARTICLE 388
SURFACE METAL RACEWAYS – ARTICLE 386
SURFACE NONMETALLIC RACEWAYS – ARTICLE 388
USES NOT PERMITTED • 386.12
(1) WHERE SUBJECT TO SEVERE PHYSICAL DAMAGE
(2) WHERE 300 V OR MORE BETWEEN CONDUCTORS, UNLESS METAL IS 0.040" THICK OR GREATER
(3) WHERE SUBJECT TO CORROSIVE VAPORS
(4) IN HOISTWAYS
(5) CONCEALED, EXCEPT PER 386.10(4)
COMPONENTS
1 ENTRANCE END FITTING
2 TWO-GANG DEVICE BOX
3 INTERNAL CORNER
4 EXTERNAL CORNER
5 SINGLE-GANG DEVICE BOX
6 PLUGMOLD MULTIOUTLET SYSTEM
7 COVER CLIP
8 SIZE REDUCING CONNECTOR
9 BLANK END FITTING

Figure 2-4. Chapter 3, the "Build" chapter, includes wiring methods and wiring materials to get electrical current from point "A" to point "B" in all electrical installations.

Chapter 4 covers equipment that uses electrical energy to perform a task, such as luminaires (lighting fixtures), appliances, air conditioners, motors, and heating, de-icing, and snow-melting equipment. In addition, Chapter 4 deals with equipment that controls or facilitates the use of electrical energy, such as switches, switchboards, panelboards, cords, cord caps, and receptacles. Chapter 4 also covers electrical equipment that generates or transforms electrical energy, such as generators, transformers, batteries, phase converters, capacitors, resistors, and reactors.

Among the key words and clues that should guide the *Codeology* user to think "Use, Chapter 4" are the following:

- Cords or cables
- Luminaires (lighting fixtures and fixture wires)
- Receptacles and switches
- Switchboards and panelboards
- Appliances
- Heating equipment
- Air-conditioning equipment
- Motors
- Batteries
- Generators

For additional information, visit qr.njatcdb.org Item #1064

- Transformers
- Phase converters
- Capacitors, resistors, and reactors
- Storage batteries

Specials

When a question or need for information arises dealing with special occupancies, equipment, or conditions, think "Special" and start in Chapters 5, 6, or 7. **See Figure 2-5.** Chapters 1 through 4 are the foundation or backbone for all electrical installations. For installations that contain a "Special Occupancy, Equipment, or Condition," Chapters 5, 6, or 7 supplement or modify the first four chapters to address the "Special" situation.

Key words and clues that should steer the *Codeology* user to think "Special Occupancies, Chapter 5" include the following:

- Hazardous, classified locations
- Commercial garages, motor fuel dispensers
- Spray booths and applications
- Hospitals and all health care facilities
- Places of assembly
- Theaters
- Carnivals
- Temporary power
- Manufactured buildings, motor homes, RVs
- Agricultural buildings, farms
- Marinas, floating buildings

Key words and clues that should guide the *Codeology* user to think "Special Equipment, Chapter 6" include the following:

- Electric signs
- Cranes
- Elevators
- Electric welders
- X-ray equipment
- Swimming pools
- Solar Photo Voltaic Systems
- Fuel cell systems
- Wind Electric Systems
- Fire pumps

Key words and clues that should guide the *Codeology* user to think "Special Conditions, Chapter 7" include the following:

- Emergency systems
- Legally required and optional standby systems
- Systems with Class 1, 2 and 3 power sources
- Fire alarm systems
- Fiber optics

Communication

When a question or need for information in the *NEC* deals with communications systems, think "Communication" and start in Chapter 8. Chapter 8 stands alone in that the rest of the *NEC* does not apply to this chapter unless specifically referenced as such within Chapter 8. **See Figure 2-6.** Key words and clues that should guide the *Codeology* user to think

Figure 2-5 "Specials"

Figure 2-5. Chapters 5, 6, and 7 of the NEC *are the "Special" chapters. Chapter 5 covers Special Occupancies (including hazardous locations), Chapter 6 covers Special Equipment (including spas), and Chapter 7 covers Special Conditions (including emergency systems).*

Figure 2-6 "Communication"

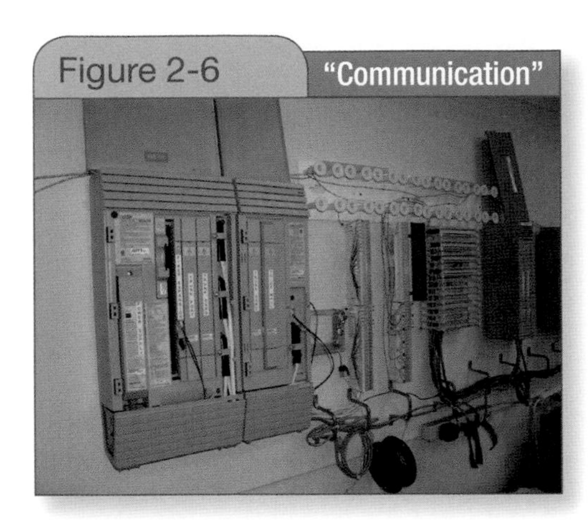

Figure 2-6. Chapter 8 of the NEC *covers communications systems.*

"Communications, Chapter 8" include the following:

- Communications circuits
- Radio and TV
- CATV systems
- Network-powered broadband systems

The *Codeology* outline provides a simplified view of all *NEC* chapters and how they are placed in specific groups when using the *Codeology* method. **See Figure 2-7.**

Fundamental Steps to Using *Codeology*

When a need arises to find information in the *NEC*, several steps may be necessary to find the answers. The following are situations or circumstances for which such information may be sought:

- Job site problems/questions
- Design concerns
- Inspection questions
- *NEC* proficiency exam

The following steps should be followed to answer any problems/questions:

Step 1

- Qualify the question or need
- Go to the Contents pages

Figure 2-7 The *Codeology* Method

THE GROUND RULES		
Introduction	Article 90	Introduction/Directions
THE BASIC INSTALLATION		
Chapter 1	100 Series	General
Chapter 2	200 Series	Plan
Chapter 3	300 Series	Build
Chapter 4	400 Series	Use
THE SPECIALS		
Chapter 5	500 Series	Occupancies
Chapter 6	600 Series	Equipment
Chapter 7	700 Series	Conditions
COMMUNICATIONS: THE LONER		
Chapter 8	800 Series	Communications
TABLES		
Chapter 9	Tables and Annexes	

Figure 2-7. The Codeology *method is broken down into five groups to aid the* Code *user in finding information in the* NEC.

- Look for clues or key words to go to the correct chapter
- Get to the correct chapter

Step 2
- Further qualify the question or need
- Look for clues or key words
- Use the Contents pages to get to the correct article within the correct chapter

Step 3
- Further qualify the question or need
- Look for clues or key words
- Use the Contents pages to go into the correct part of the correct article

Step 4
- Open the *NEC* to the part of the article that meets the question or need

Step 5
- Read only the section titles to find the correct section

Step 6
- Read section title and all titles of first-level subdivisions

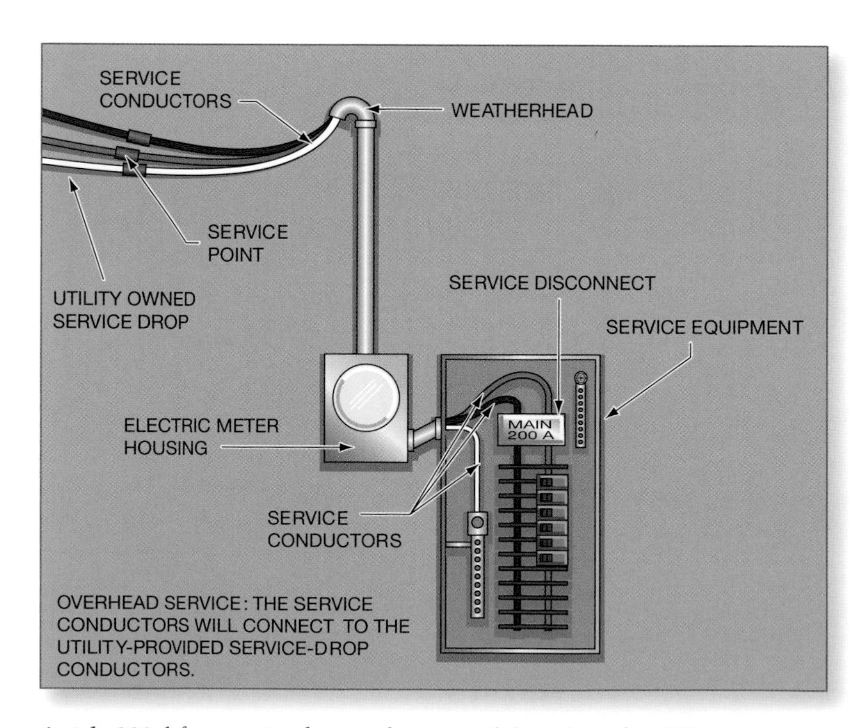

Article 100 defines service drop conductors used throughout the NEC.

- Read all of the section and pertinent subdivisions including exceptions and Informational Notes
- Apply the rule or answer the question

APPLYING THE *CODEOLOGY* METHOD – *NEC* CHAPTER 1, "GENERAL"

Chapter 1 comprises two articles, one providing the definitions essential for the proper application of the *NEC* (**Article 100**), and the second providing the general requirements for all electrical installations (**Article 110**), which is broken down into five parts.

Article 100 contains only those definitions essential to the proper application of the *Code*. It is not intended to include commonly-defined general terms or commonly-defined technical terms from related codes and standards. In general, only those terms that are used in two or more articles are defined in **Article 100**. Other definitions are included in the article in which they are used, but may be referenced in **Article 100**.

Article 100 contains definitions for terms that are not commonly used or are essential to the proper application of the *NEC*. A term that needs to be defined in the *NEC* is placed in **Article 100** only if it is used in two or more articles. Terms that are defined in the *NEC* and are used in only one article are defined in the second section of the article in which they appear. For example, in **Article 517 Health Care Facilities**, forty-seven definitions are listed in **Section 517.2**, the second section of the article. When terms are defined within an individual article, they apply only within that article.

Part I of **Article 100** contains definitions intended to apply wherever the terms are used throughout the *Code*. **Part II** contains five definitions applicable only to the parts of articles specifically covering installations and equipment operating at over 1000 volts, nominal. For

Figure 2-8 Layout of *NEC* Chapter 1

***NEC* Title: General** ***Codeology* Title: General** **Chapter Scope: General Information and Rules for Electrical Installations**	
Article	**Title**
100	Definitions
110	Requirements for Electrical Installations

Figure 2-8. Article 100 and Article 110 of Chapter 1 apply to all electrical installations unless altered by Code *language of Chapters 5, 6, and 7.*

example, terms defined in **Part II** of **Article 100** would apply to **Part IX** of **Article 240. See Figure 2-8.**

Misapplication of the *Code* requirements often occurs due to a lack of understanding of terms defined in **Article 100**. For instance, to understand and apply the requirements of **Article 230 Services**, the user must understand the terms defined in **Article 100**, including but not limited to:

Service. The conductors and equipment for delivering electric energy from the serving utility to the wiring system of the premises served.

Service Cable. Service conductors made up in the form of a cable.

Service Conductors. The conductors from the service point to the service disconnecting means.

Service Conductors, Overhead. The overhead conductors between the service point and the first point of connection to the service-entrance conductors at the building or other structure.

Service Conductors, Underground. The underground conductors between the service point and the first point of connection to the service-entrance conductors in a terminal box, meter, or other enclosure, inside or outside the building wall.

Informational Note: Where there is no terminal box, meter, or other enclosure, the point of connection is considered to be the point of entrance of the service conductors into the building.

Service Drop. The overhead conductors between the utility electric supply system and the service point.

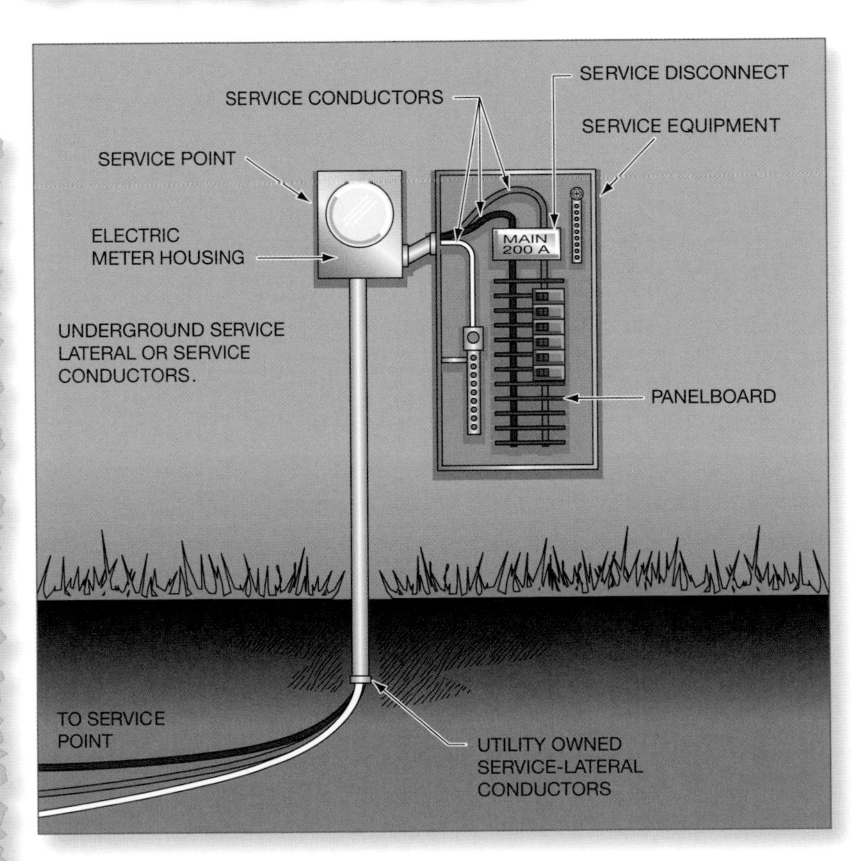

Article 100 defines underground service lateral conductors used throughout the NEC.

Service-Entrance Conductors, Overhead System. The service conductors between the terminals of the service equipment and a point usually outside the building, clear of building walls, where joined by tap or splice to the service drop or overhead service conductors.

Service-Entrance Conductors, Underground System. The service conductors between the terminals of the service equipment and the point of connection to the service lateral or underground service conductors.

Informational Note: Where service equipment is located outside the building walls, there may be no service-entrance conductors or they may be entirely outside the building.

Service Equipment. The necessary equipment, usually consisting of a circuit breaker(s) or switch(es) and fuse(s) and their accessories, connected to the load end of service conductors to a building or other structure, or an otherwise designated area, and intended to constitute the main control and cutoff of the supply.

Service Lateral. The underground conductors between the utility electric supply system and the service point.

Service Point. The point of connection between the facilities of the serving utility and the premises wiring.

Informational Note: The service point can be described as the point of demarcation between where the serving utility ends and the premises wiring begins. The serving utility generally specifies the location of the service point based on the conditions of service.

FACT

Due to the different characteristics of conductor metals, all terminals, different connectors, and lugs shall be identified (labeled/marked) for the conductor material permitted. Dissimilar metals, such as aluminum and copper, shall not be intermixed in a connector unless the connector is identified for the purpose. Solder, flux, inhibitors, and conductor compounds must not have an adverse effect on the conductors or equipment.

The term "service" is actually part of various definitions of **Article 100** other than the ones above: Bonding Jumper, Main; Feeder; Listed; Overcurrent Protective Device, Branch-Circuit; Premises Wiring (System); Separately Derived System; and Surge-Protective Device (SPD). **Article 100** contains many additional definitions that are essential to the proper application of **Article 230**. However, up to eleven separate definitions may be needed to properly identify conductors and equipment in order to ensure proper application of **Article 230**. Before interpreting and applying the rules of **Article 230** or any other article, the *Code* user must understand the definitions that apply.

ARTICLE 110 REQUIREMENTS FOR ELECTRICAL INSTALLATIONS

Article 110 contains general requirements, separated into five parts, which are common to all electrical installations. **See Figure 2-9.** The key provisions of this article represent the general nature of these requirements as they apply to all electrical installations.

- General
- **Section 110.1 Scope**
- Examination and approval of conductors and equipment
- Installation and use of conductors and equipment
- Access to electrical equipment
- Spaces about electrical equipment
- Enclosures intended for personnel entry
- Tunnel installations

In addition to the common requirements listed in this section, a complete review of all the *Code* language of **Article 110** is very important. Refer to the *NEC* and read the following sections:

- **110.2** Approval occurs when the conductors and equipment are acceptable to the authority having jurisdiction (AHJ).
- **110.3** Examination, identification, installation and use of equipment are identified in the eight rules used to approve installations.
- **110.3(B)** Listed or labeled equipment is required to be installed in accordance with the listing and labeling of the equipment. **See Figure 2-10.**
- **110.5** Conductors referenced in the *NEC* shall be copper or aluminum

Figure 2-9 Five Parts of Article 110

Part I.	General
Part II.	1000 Volts, Nominal, or Less
Part III.	Over 1000 Volts, Nominal
Part IV.	Tunnel Installations over 1000 Volts, Nominal
Part V.	Manholes and Other Electric Enclosures Intended for Personnel Entry, All Voltages

Figure 2-9. Article 110 is divided into five parts that identify the basic requirements for electrical installations.

unless otherwise provided in the *Code*. When the conductor material is not specified in a *Code* section, the material and sizes given apply to copper conductors.

- **110.6** Conductor sizes used in the *NEC* are expressed in American Wire Gage (AWG) or in circular mils. See Table 8 in Chapter 9 of the *NEC* showing the circular mil sizes for conductors using the AWG.
- **110.9** All equipment intended to interrupt current at fault levels must have a sufficient interrupting rating.
- **110.10** Damage to electrical systems from short circuits and ground faults must be minimized through proper selection of protective equipment and

Figure 2-10 UL Listed Labels

Figure 2-10. Underwriters Laboratories' first test was in 1895 under the name of Underwriters' Electrical Bureau. Today, the electrical equipment installed is listed in the UL White Book.

For additional information, visit qr.njatcdb.org Item #1542

consideration of all characteristics of the equipment and conductors protected.

- **110.11** Equipment that is not identified for use outdoors or identified for indoor use shall be protected from damage from the weather during construction. Plastic sheeting or other means for protection is commonly used to prevent this damage.
- **110.12** This section requires that all electrical equipment be installed in a neat and workmanlike manner. Electrical installations must be installed neatly and with care to be considered installed in a skillful or workmanlike manner.
- **110.12(A)** All unused cable or raceway openings (holes) in all electrical equipment must be effectively closed. Holes intended for the operation of equipment or mounting purposes do not have to be closed.
- **110.12(B)** All electrical equipment must be kept clean and undamaged by foreign materials such as paint, plaster, or other corrosive materials. Equipment must be protected from damage, corrosion, overheating, or deterioration.
- **110.13** All electrical equipment must be firmly secured to the surface on which it is mounted.
- **110.13(B)** Electrical equipment that depends on the natural circulation of air for cooling purposes must be installed such that other equipment, walls, or other structural members do not block or limit the natural circulation of air.
- **110.14** All electrical installations contain electrical connections. This section contains general requirements for all electrical connections in all electrical installations.
- **110.14(A)** Conductors must be terminated in such a manner to ensure a solid connection without damaging the conductors. Terminals designed for the termination of more than one conductor must be identified for such use. Terminals designed for use with aluminum conductors must also be identified for the purpose.
- **110.14(B)** Conductors are required to be spliced using approved methods.
- **110.14(C)** All electrical terminations and conductors are temperature limited, typically at 60°C, 75°C, and 90°C. The intent of this requirement is to ensure that the lowest temperature rating in an electrical system or circuit is not exceeded.
- **110.15** The use of a 3-phase, 4-wire delta-connected system results in one phase having a higher voltage to ground. For example, a 120/240 volt, 3-phase, 4-wire system is required, by **Article 250** of the *NEC*, to be grounded. **See Figure 2-11.**

FACT

Wooden plugs are prohibited for mounting or securing electrical equipment. In the event of a fire, wooden plugs fail quickly and the equipment may fall, causing injury to those exiting from the location.

Figure 2-11 **Delta High Leg**

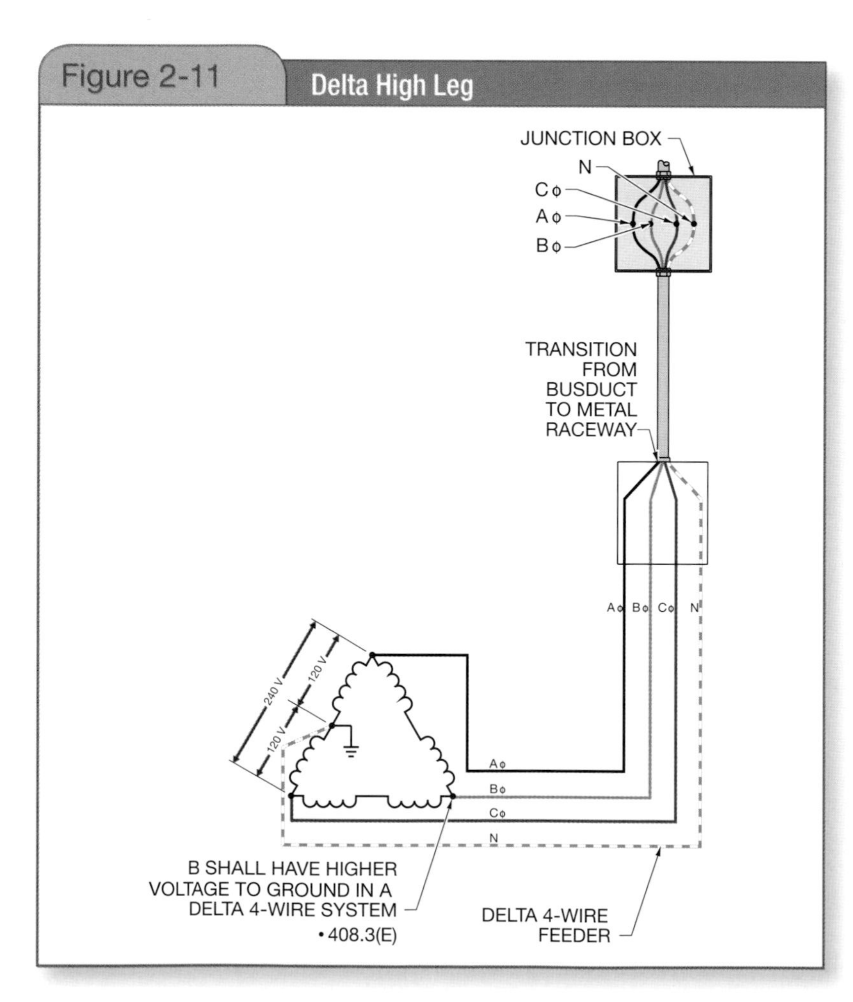

Figure 2-11. The high leg, often called the wild leg, must be installed as the "B" phase on a 4-wire delta system and identified with the color orange.

An arc flash can produce temperatures of 35,000°F at the point of contact, and temperatures in the ambient space of an Electrical Worker can easily reach upwards of 15,000°F. Accidents involving an arc flash can result in serious injuries or death from incurable third-degree burns. The marking requirement in Section 110.16 is designed to warn qualified persons of potential arc flash hazards. Arc flash warnings are required to be marked on all switchboards, panelboards, industrial control panels, meter socket enclosures, and motor control centers used in locations other than dwelling units. An informational note informs the *Code* user that *NFPA 70E: Standard for Electrical Safety in the Workplace* provides guidance in determining the severity of an exposure, safe work practices, and required personal protective equipment.

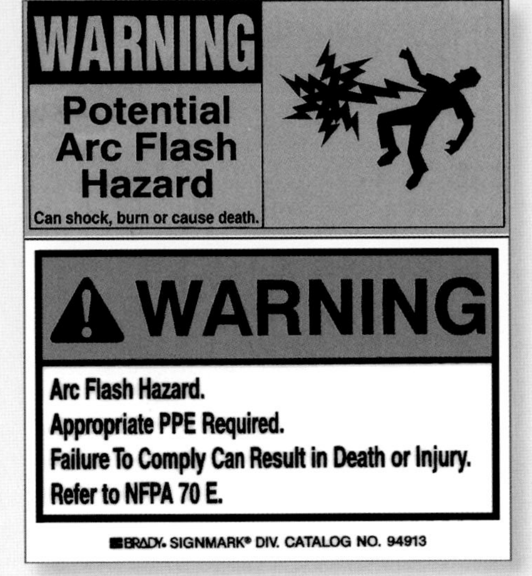

Personal protective equipment is available in various calorie ratings. NFPA 70E is a valuable resource to be studied for arc flash protection.

- **Section 110.16 Arc Flash Hazards** and labeling requirements to warn people of the potential hazards due to arc flash or arc blast.
- **110.21** All electrical equipment is required to be durably marked with the manufacturer's name, trademark, or other distinctive marking for identification purposes. Field-applied hazard markings meeting the ANSI Z235.4-2011 standard are also required.
- **110.22** All disconnecting means are required to be legibly marked to identify their purpose, unless the installation is so arranged that the purpose of the disconnecting means is clearly evident.
- **110.24** Service equipment, other than dwelling units, shall be marked legibly in the field listing the maximum available fault current. This marking shall include the date the fault current was calculated. Upstream equipment information such as transformer size, distance to the next distribution equipment or transformer, motor loads, and other factors are all part of the calculation for available fault current.

Part II. 1000 Volts, Nominal, or Less

- **110.26** This section requires that sufficient access to equipment and working space be maintained. All electrical equipment likely to require examination, adjustment, servicing, or maintenance while energized must provide adequate working space.
- **110.26(A)** Adequate working space is required for the safety of electrical workers who install and maintain electrical systems.

Part IV. Tunnel Installations over 1000 Volts, Nominal

Part IV of **Article 110** is dedicated to tunnel installations with systems and circuits rated over 1000 volts. **Section 110.51** explains in detail the areas

FACT

The following are designed for wire splicing:
- Devices identified for the purpose (for example, wire nuts)
- Brazing, welding, or soldering with a fusible metal or alloy

Soldered spliced must first be mechanically joined, and all splices and free ends of conductors must be insulated. Wire nuts or splicing devices for direct burial must be listed for the purpose.

DANGER indicates a hazardous situation which, if not avoided, WILL result in DEATH or SERIOUS INJURY.

WARNING indicates a hazardous situation which, if not avoided, COULD result in DEATH or SERIOUS INJURY.

CAUTION indicates a hazardous situation which, if not avoided, MAY result in MINOR or MODERATE INJURY.

The words "danger," "warning," and "caution" are used in the NEC. These signal words indicate the degree of hazard. Other label components provide specific instruction regarding hazards and can also have a graphic included in the instruction, showing the hazard.

covered by **Part IV.** This part covers the installation and use of high-voltage power distribution and utilization equipment which is portable, mobile, or both. Examples include, but are not limited to, substations, trailers, cars, mobile shovels, draglines, hoists, drills, dredges, compressors, pumps, conveyors, and underground excavators.

Part V. Manholes and Other Electrical Enclosures Intended for Personnel Entry

Part V of **Article 110** is dedicated to manholes and other electric enclosures intended for personnel entry at all voltages. It addresses working space inside these enclosures and the provision of adequate access and space where equipment or parts contained are likely to require examination, adjustment, servicing, or maintenance while energized. Enclosures must also be of sufficient size to allow for the installation and removal of conductors without damaging the conductor insulation.

(1) Depth of working space. Table 110.26(A)(1) Working Spaces specifies the minimum working space in front of electrical equipment.

This table classifies systems of 1000 volts or less in either of two categories using the voltage to ground system: 0 to 150 volts to ground, 151 to 600 volts to ground, and 601 to 1000 volts to ground. Minimum clear distance for working space is then determined from one of three conditions:

Condition 1 – The equipment is opposite no other electrical equipment or any grounded objects or parts.
Condition 2 – The equipment is opposite grounded objects or parts. Note that concrete, brick, or tile walls are considered grounded.
Condition 3 – The equipment is opposite other electrical equipment.

(2) Width of working space. The width of working space in front of all electrical equipment is required to be the width of the equipment or 30 inches, whichever is greater. In all cases, hinged panels and doors must be capable of opening a full 90°.

(3) Height of working space. The working space must be clear and extend from the floor or working platform to the height of the equipment or 6½ feet, whichever is greater.

110.26(B) Clear Spaces – The working space in front of electrical equipment must not be used for storage.

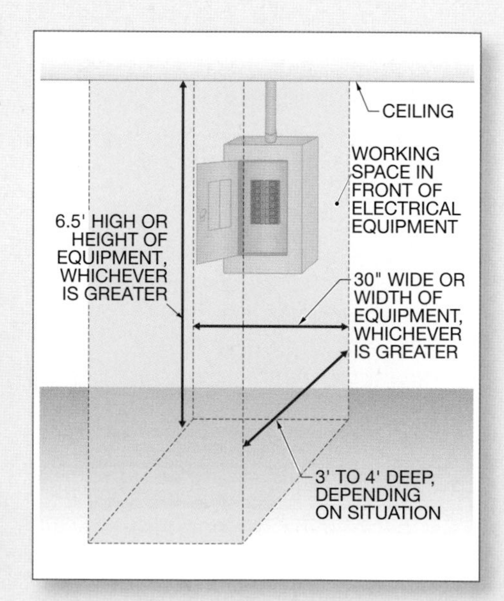

Nothing can be stored in the workspace area. This is to ensure that personnel can access the electrical equipment quickly.

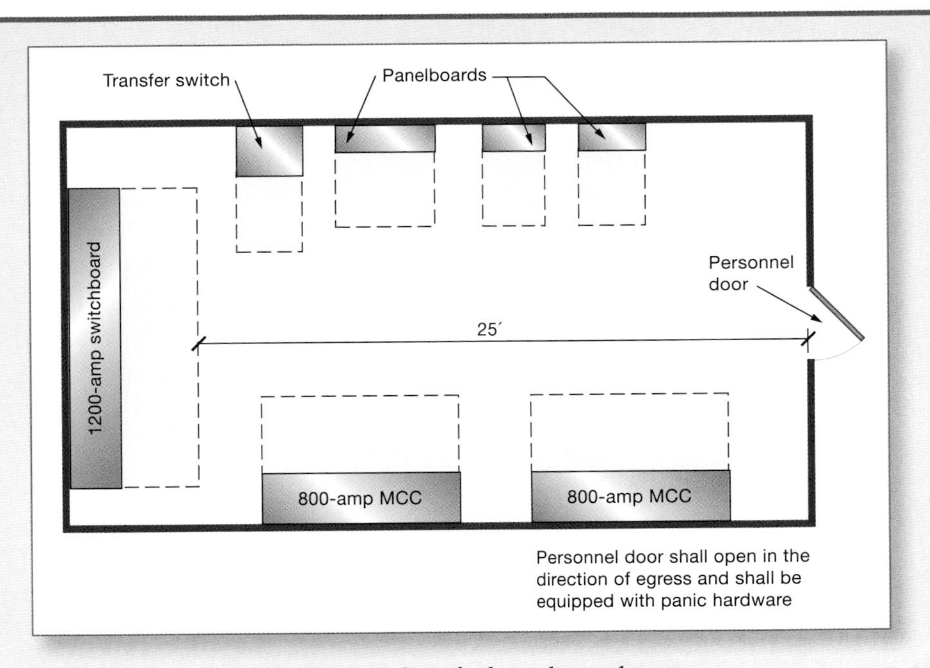

The NEC *requires panic hardware on doors for large electrical rooms.*

110.26(C) (3) Personnel Doors. Where equipment rated 800 amps or more that contains overcurrent, switching, or control devices is installed, all personnel doors less than 25 feet from the nearest edge of the working space must open in the direction of egress and be equipped with listed panic-type hardware to permit a quick exit in the event of an electrical fault.

110.26(D) Illumination – Working spaces around service equipment, switchboards, switchgear, panelboards, or motor control centers installed indoors require illumination to provide Electrical Workers maintaining the system with adequate lighting to perform routine tasks.

110.26(E)(1)(a) Dedicated Electrical Space – This section applies to all switchboards, panelboards, and motor control centers, and requires that the locations where they are installed are dedicated to this equipment. This requirement is not working space for the Electrical Worker; rather it is dedicated equipment space to allow for raceways and cable assemblies to enter equipment. This requirement is also intended to prevent foreign systems from being installed above electrical equipment. The dedicated space addressed in this section is the area above the footprint formed by the top of the equipment.

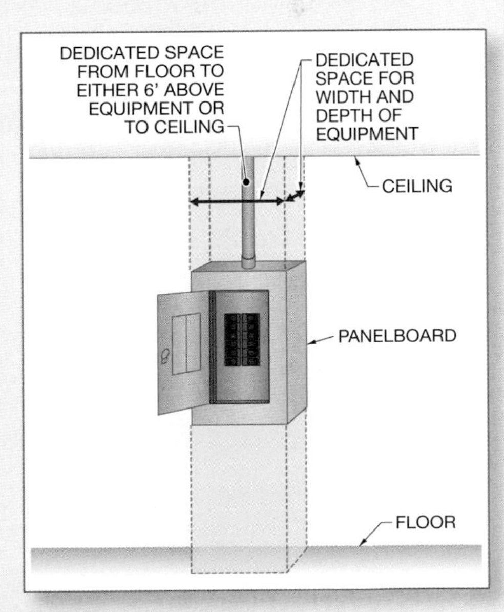

No piping, ducts, leak protection apparatus, or other equipment foreign to the electrical installation shall be located in the dedicated space

KEY WORDS AND CLUES FOR CHAPTER 1, "GENERAL"

- General definitions for terms used in more than one article
- Determination of approval
- Examination of equipment
- Identification of equipment
- Installation and use of equipment
- General questions about electrical installations
- Mechanical execution of work
- Mounting and cooling of equipment
- Electrical connections
- High-leg marking requirements
- Flash protection
- Markings
- General identification of disconnects
- General electrical questions for over 1000 volts
- General questions for manholes and tunnels
- Working space

Summary

Learning, understanding, and applying the *Codeology* method begins with four basic building blocks:

1. **Contents** pages: The *Code* user must be familiar with and understand the 10 major subdivisions of the *NEC*.
2. **Arrangement** of the *NEC* as detailed in **Section 90.3**: The *Code* user must understand how the individual chapters of the *NEC* apply.
3. Structure or **Outline** form of the *NEC*: Without an understanding of the hierarchy of requirements and information in the *NEC*, proper application would be impossible.
4. **Language** of the *NEC*: An in-depth understanding of defined terms is essential for proper application of *NEC* requirements.

The *Codeology* outline builds upon the arrangement of the *NEC* and provides a user-friendly method to help categorize or qualify a question or need through key words and clues.

Understanding the *Codeology* outline along with a solid foundation in the four basic building blocks allows the *Code* user to begin to apply the *Codeology* method. Additional fundamentals include recognizing key words and clues to provide a quicker and accurate method for discovery of needed information. All of these fundamentals must be solidly in place before the *Codeology* method is applied to the introduction and nine chapters of the *NEC*.

Summary

In accordance with **Section 90.3**, Chapter 1 applies generally to all electrical installations. Using the *Codeology* method, this chapter has been called the "General" chapter, due to the breadth of its coverage. The scope of this chapter, which can be described as "General Information and Rules for all Electrical Installations," covers the entire electrical system, from the service point (connection to the utility) to the last receptacle or outlet in the electrical system.

Chapter 1, along with Chapters 2, 3, and 4, builds the foundation or backbone of all electrical installations.

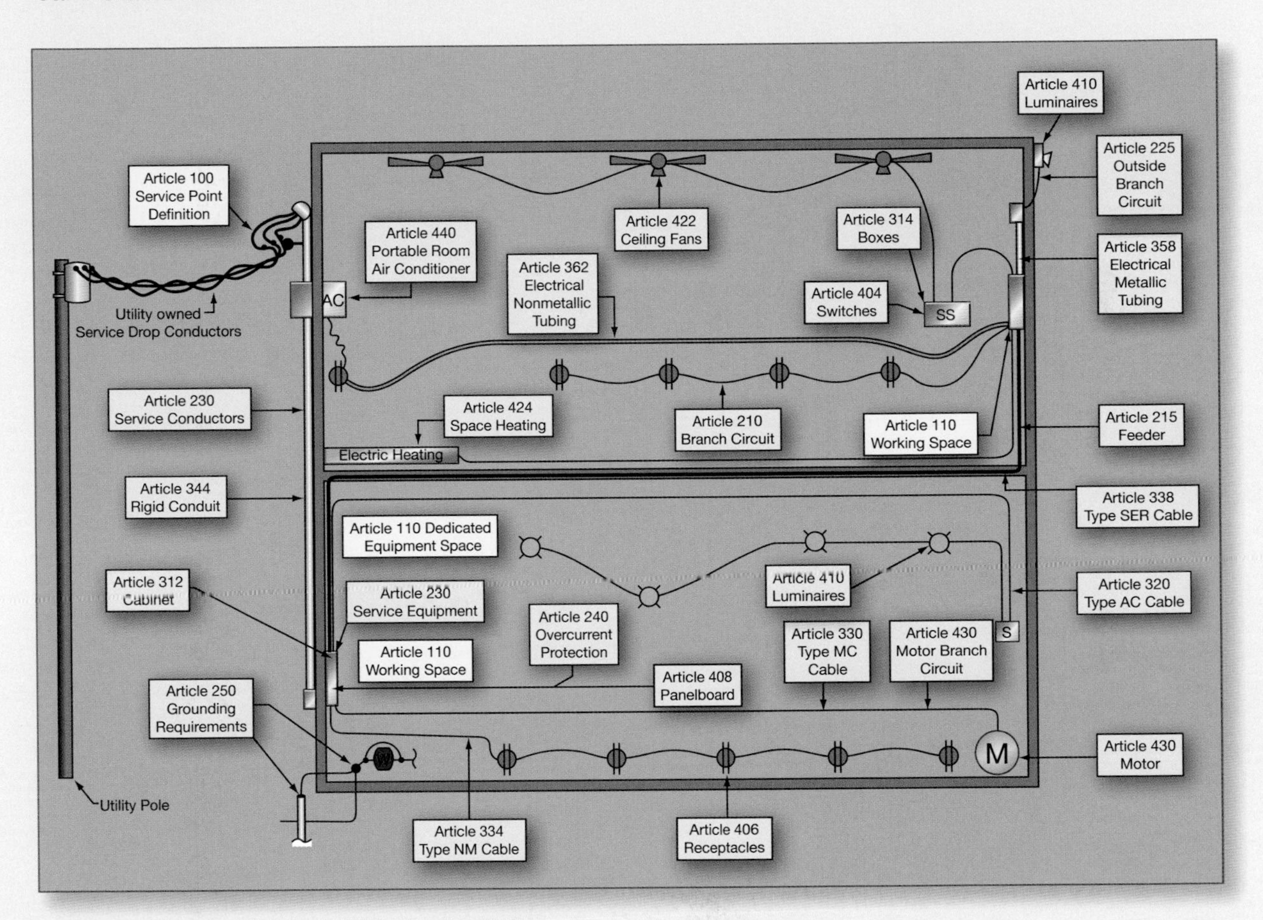

Chapter 1 applies generally in all electrical installations.

Review Questions

1. **Does Chapter 1 of the *NEC* apply to all electrical installations covered by the *Code*?**
   a. Yes
   b. No
   c. Only when applied and used for installations in Chapter 1
   d. Only when applied in Chapters 1–4
2. **Chapter 1 is divided into __?__ articles.**
   a. 2
   b. 3
   c. 8
   d. 9
3. **Article 100 Definitions contains all terms defined in the *NEC*.**
   a. True
   b. False
4. **When is a definition placed in Article 100 of the *NEC*?**
   a. Only when it is a commonly used term
   b. Upon the consensus of all of *Code*-Making Panels
   c. When the term does not qualify for inclusion in an article
   d. When the term is used in two or more articles
5. **When a term is defined in an article other than Article 100, which section is it placed in?**
   a. XXX.2
   b. XXX.3
   c. XXX.4
   d. XXX.5
6. **The provisions of Part II of Article 100 apply to the installation of a 120/240 volt panelboard in a dwelling unit.**
   a. True
   b. False
7. **Workspace clearances are general requirements for all electrical installations. Where in Chapter 1 is this general requirement for equipment rated at 240 volts located?**
   a. **Part I**
   b. **Part II**
   c. **Part III**
   d. **Part IV**
8. **Article 110 is divided into __?__ parts.**
   a. 3
   b. 4
   c. 5
   d. 6

Review Questions

9. Part IV of Article 110 is limited to installations above 1000 volts.

a. True

b. False

10. __?__ of Article 110 would contain the requirement that all listed and labeled equipment be installed in accordance with their listing and labeling.

a. **Part I**

b. **Part II**

c. **Part III**

d. **Part IV**

11. Using the *Codeology* method, *Code* users must determine the proper clues or key words and go to the __?__.

a. Contents pages

b. highlighted areas

c. index

d. most likely article

12. After using key words or clues to get into the correct chapter of the *NEC*, *Codeology* users must further qualify their needs and go to the correct __?__.

a. article

b. part

c. section

d. subdivision

13. After using key words or clues to get to the correct chapter and then the correct article of the *NEC*, *Codeology* users must further qualify their needs and go to the correct __?__.

a. part of the article

b. section of the part

c. subdivision

d. table

14. Which chapter of the *NEC* is considered the "stand-alone" one?

a. 7

b. 8

c. 9

d. 10

15. An article in the 600 series will deal with special __?__.

a. equipment

b. occupancies

c. permission

d. none of the above

16. The *Codeology* method will allow the *Code* user to become __?__ with the *NEC*.

a. accurate

b. confident

c. fast

d. all of the above

The *NEC* "Plan" Chapter (Wiring and Protection)

Introduction

The *Codeology* title for Chapter 2 of the *National Electrical Code* is "Plan." This chapter contains information vital to all *Code* users, as the electrical concepts discussed are used in everyday design, construction, and maintenance. The chapter covers three wiring systems addressed throughout the *Code* and is common to all occupancies, branch circuits, feeders, and services. Also covered are important basic safety topics such as overcurrent protection for conductors and equipment, grounding and bonding, and surge suppression devices. Chapter 2 and Chapter 3 of the *NEC* are considered the "heart" or backbone of the *NEC*. When these chapters are used together, they cover the basic installation rules used for operational electrical distribution and power systems. It is very important for the user to know the content of these chapters well. The requirements in Chapter 2 and Chapter 3 serve as a foundational knowledge base for learning other vital *Code* concepts. Learning these concepts will make the rest of the *Code* easier to understand. If the user is willing to read each article completely and study the contents, a solid foundation for using the *Code* will be established.

Just as an architect or designer needs to know basic information for planning the construction of a new building or structure, users of the *NEC* must have basic knowledge of electrical installation rules. All electrical installations must be planned to suit the many needs required in the performance of an electrical system, while taking into account the total cost of materials and installation.

The minimum needs to accomplish this goal include:

- Design: electrical installation must suit the specific needs of the occupancy, building, structure, or location in which it is installed.
- Compliancy: installation practices must fall within the requirements of the *NEC* and any local building codes requirements if applicable.
- Plan submission: Conceptual and final drawings must be submitted to the authority having jurisdiction (AHJ) for final plan review and approval before starting construction.
- Planning Stages: The planning stage of all electrical installations includes two very important areas covered in Chapter 2 of the *NEC* – Wiring and Protection.

Chapter 3

Objectives

- Recite from memory the articles and their titles that form *NEC* Chapter 2.
- Use the *Codeology* method to find information in *NEC* Chapter 2.
- Recite the basic use of the *Codeology* method using the proper generic terms for each chapter.
- Demonstrate the skill of using the Table of Contents to find *Code* information.
- Associate the *Codeology* title for *NEC* Chapter 2 as "Plan."
- Identify which articles in Chapter 2 apply to "Wiring."
- Identify which articles in Chapter 2 apply to "Protection."
- Describe the "planning" type of information and requirements necessary for wiring and protection.
- Recognize Chapter 2 numbering as the 200-series.
- Recognize, recall, and apply the articles contained in Chapter 2.

Table of Contents

USING *CODEOLOGY* TO GET TO THE CORRECT CHAPTER, ARTICLE, AND PART

The *Codeology* method is designed to provide the *NEC* user with an organized and logical scheme to find answers to *Code* questions at a determined location. This method begins by using clues and key words to navigate the Table of Contents in the *NEC* and then identify the correct chapter and part for beginning the inquiry. These concepts are the fundamentals of the *Codeology* method.

The *NEC* is written in an outline format. This means that each chapter is subdivided into articles, which are subdivided into parts, which are subdivided into sections, etc. The *Codeology* method is designed to provide a disciplined, systematic approach to quickly find information by understanding and applying this outline format used in the *NEC*. The basic plan of *Codeology* is for the user to obtain a foundational knowledge of the contents and location of applicable *NEC* sections throughout the *Code* book.

Generic terms like PLAN-BUILD-USE are the basic tools for using clues and key words, along with the Table of Contents, to locate answers to questions. The generic terms used in the *Codeology* method for each chapter are listed below:

- Basic/General
- Plan
- Build
- Use
- The Specials
- Communications
- Tables and Examples

The key to success for applying the *Codeology* method is using these generic terms to classify inquiries. The question is first evaluated and classified into one of the generic term categories, which then relates to a chapter in the Table of Contents. Remember that the articles of Chapters 1 through 4 are required for all electrical installations, unless modified by Chapters 5, 6, or 7. The *Codeology* generic terms match the Table of Contents as follows:

- Basic/General – **Article 90** and Chapter 1. This information is basic or "General" in nature for all electrical installations. Chapter 1 begins with **Article 90**, which is considered a preface to the *NEC*. It covers the purpose of the *Code*, what occupancies are (and are not) covered by the *NEC*, arrangement and layout, enforcement, language, and equipment safety. Chapter 1 also has two Articles – Definitions and Requirements for Electrical Installations – that cover general installation requirements intended for all electrical installations including equipment examination, approval and requirement considerations, installation techniques, mechanical execution of work, electrical connections, safety, and working clearances.
- Plan – Chapter 2 of the *NEC*. This term is used for determining the planning stages of electrical work and is general in nature as it applies to all installations. Chapter 2 covers two subjects: wiring and protection. The wiring section covers grounded (sometimes called neutral) conductors, branch circuits, feeders installed both inside and outside a building, calculations, and services. The protection section of Chapter 2 covers overcurrent protection devices, grounding and bonding, and surge protection. This chapter is fairly diverse and is a key part of *Code* installations. Every circuit conductor is classified as a branch circuit, feeder, or service. Every *Code* user should spend time becoming familiar with the contents, as they are very common rules for all installations.

- Build – Chapter 3 of the *NEC*. This chapter covers wiring methods and materials installed during construction. General installation rules, enclosures, boxes, cables, raceways, and open wiring on insulators are covered here. Any question that specifically addresses wiring and materials can be found here. Use the Table of Contents to locate specific information.
- Use – Chapter 4. This chapter covers cords, devices, and equipment generally used after the electrical installation is complete. For example, **Article 404** covers Switches, **406** covers Receptacles, **422** Appliances, **430** Motors, etc. Questions can be easily qualified to this chapter by content.
- The Specials – Chapters 5, 6, and 7. These chapters address special occupancies, equipment, or conditions. These articles are for very specific installations. They can change the installation requirements found in Chapters 1 through 7, sometimes requiring strict installation criteria for safety. In other words, they supplement or modify the first seven chapters of the *Code* for "Special" installations due to occupancy use, equipment, and special systems.
- Communications – Chapter 8. This chapter is a "stand alone" chapter, meaning it does not apply to the other seven chapters of the *Code* unless specifically referenced. This chapter addresses communication circuits, radio and TV systems, and network broadband systems.
- Tables and Informative Annex – Chapter 9 is used for several calculations required by *NEC* sections. The Informative Annex located at the end of the book is used to provide helpful information, calculation examples, and other information needed by a user but not required for enforcement of the *Code*.

APPLICATION OF THE *CODEOLOGY* METHOD

With practice, the *Codeology* method is quite easy to use. The key to success in using the *Codeology* method is to know, or even memorize, the article numbers and their topical designations. Commit the following to memory from Chapter 2:

- **200** – Use and Identification of Grounded Conductors
- **210** – Branch Circuits
- **215** – Feeders
- **220** – Branch Circuit, Feeder, and Service Calculations
- **225** – Outside Branch Circuits and Feeders
- **230** – Services
- **240** – Overcurrent Protection
- **250** – Grounding and Bonding

An example of the process for using the *Codeology* method is as follows. The question is asked, "Can fuses or circuit breakers be installed in a dwelling unit bathroom?" Using the index to locate the answer to this question is fairly difficult, so it is best to use the *Codeology* method to find the answer. Seeking a clue or key word to begin the process, first determine that the question is asking for the location of "fuses and circuit breakers," which is installation information. This is a "Plan" section question (Chapter 2 in the *NEC*), but due to the fact it is very generic in nature, the next step is to look at the Table of Contents.

Article 240 addresses the subject of Overcurrent Protection, which is what fuses and circuit breakers are, and **Part II** of **Article 240** covers "Location" of overcurrent devices. **240.21** is the first article regarding location.

Now scan the pages reading the bold section titles only, looking for a clue to answer the question. **240.24** contains the clue to answer the question, **240.24(E)** Not Located in Bathrooms. Reading the section overcurrent devices shall not be installed in dwelling unit bathrooms is found.

The following practice problems are designed to demonstrate the use of the *Codeology* method for finding information in Chapter 2, "Plan" section of the *NEC*.

Question: In a premises wiring system, *branch circuits* supplied by a system of *more than one voltage*, each branch-circuit ungrounded conductor must be identified by what means?

Step 1. Identify the key words and clues in the question. They are identified above.

Step 2. Go to the Table of Contents in the *Code* book. Branch Circuits are in the "Plan" section of Chapter 2 and are part of the "Wiring" section and is identified as **Section 210**. Because the question addresses marking of conductors, use **Part I. General Provisions**. Go to the page number listed in the Table of Contents.

Step 3. Starting with **210.1**, read the **BOLD** section titles and numbers until a key word or clue is located, indicating a possible answer to the above question.

Step 4. Notice that **210.5** addresses the subject of "Identification for Branch Circuits" and item **(C)** is more specific, addressing "Identification of Ungrounded Conductors."

Step 5. Find the answer. BE SURE TO READ THE ENTIRE SECTION! **Articles 210.5(C)(1) (a)** and **(b)** answer the question.

Answer: Each branch circuit shall be identified by phase or line at all termination, connection, and splice points. The means for identification is permitted by separate color coding of each conductor.

Question: The method for equipment grounding, applied for a grounded system, shall be made by bonding the __?__ to the __?__.

Step 1. Identify the key words and clues in the question.

Step 2. Go to the Table of Contents in the *Code* book. Grounding and Bonding located in Chapter 2 in the "Protection" section and is identified as section **250.** Looking through the Parts in **250**, notice the section titled **Part VII. Methods of Equipment Grounding** fits the key words identified in the question.

Go to the page number listed in the Table of Contents.

Step 3. Starting with **Article 250.130**, note that the first section addresses "Equipment Grounding Connections." Reading the **BOLD** sections. Subsection **(A)** is for "Grounded Systems." The answer to the question can be found by reading the contents of this subsection.

Answer: The method for equipment grounding for a grounded system connection shall be made by bonding the equipment grounding conductor to the grounding electrode conductor.

CODEOLOGY DEFINITION OF "PLAN"

Chapter 2 is divided into two general areas: Wiring and Protection. These two subjects are addressed with their subcategories.

WIRING

In the planning stages of any electrical installation, the designer's intent can be seen on the electrical drawings. These drawings provide the installer with the necessary information to build the installation. All electrical installations must have a source of electrical energy – in most cases, a service (**Article 230**) – from a local utility company. From the service equipment, feeders (**Article 215**) supply panelboards to distribute electrical energy throughout the building. From the panelboards, branch circuits (**Article 210**) supply receptacle outlets, lighting fixtures, and other utilization equipment to facilitate the use of electrical energy. Branch circuits and feeders installed outside of a building or structure must also comply with outside branch circuits and feeders (**Article 225**). Additional articles within Chapter 2 are dedicated to wiring. **See Figure 3-1**.

All of the different types of conductors must be properly sized to handle the intended load. This requires that calculations (**Article 220**) be applied to properly size each conductor. When the electrical system employs a grounded conductor (**Article 200**), the installer must plan for the proper use and identification of these conductors. **See Figure 3-2**.

Code compliance for current-carrying conductors must be part of the planning process of the electrical installation. This process, covered in the wiring articles of Chapter 2, focuses only on the *NEC*

Figure 3-1 Chapter 2 Wiring (and Protection)

***NEC* Title:**	**Wiring and Protection**
***Codeology* Title:**	**Plan**
Chapter Scope:	**Information and Rules on Wiring and Protection of Electrical Installations**

WIRING

Article	Article Title
200	Use and Identification of Grounded Conductors
210	Branch Circuits
215	Feeders
220	Branch-Circuit, Feeder, and Service Load Calculations
225	Outside Branch Circuits and Feeders
230	Services

Figure 3-1. Articles ***200, 210, 215, 220, 225,*** *and* ***230*** *cover the wiring of Chapter 2.*

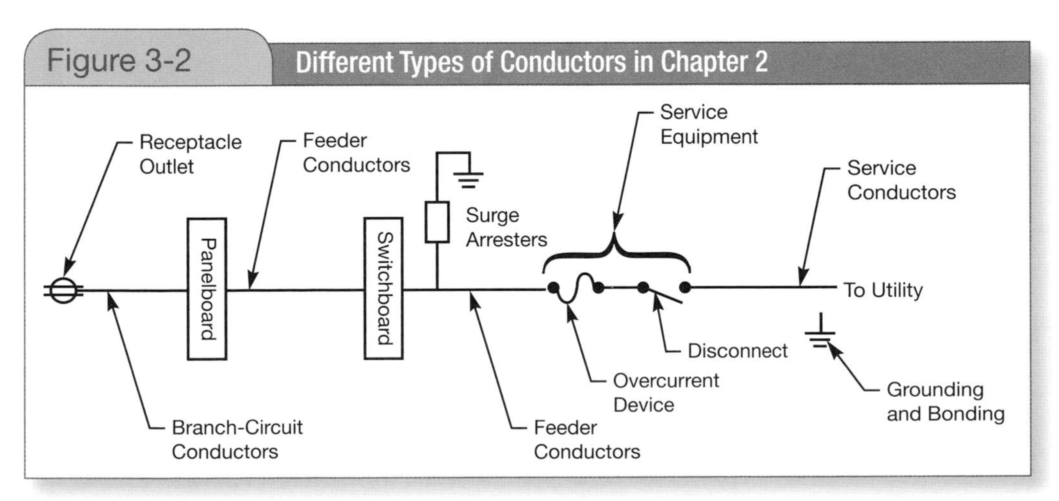

Figure 3-2. Chapter 2, the "Plan" chapter, does not address specific wiring methods, only the terms "service," "feeder," and "branch circuit."

terms for the current-carrying conductors, not the type of raceway or cable assembly. **See Figure 3-3.** These *NEC* terms for conductors are major clues for starting in Chapter 2. *NEC* terms related to planning the wiring of an electrical installation include the following:

- Branch circuits
- Feeders
- Services
- Calculations, computed load(s)
- Grounded conductors

Article 200 Use and Identification of Grounded Conductors

The term "grounded conductor" is defined in **Article 100**. **See Figure 3-4.**

Grounded conductors are often called neutral conductors. It is extremely important to note that not all grounded conductors are neutrals. When a current-carrying conductor is common to all of the ungrounded (hot) conductors of the system, it is considered neutral. When that common or neutral conductor is intentionally grounded, it becomes the grounded conductor.

Article 200 provides requirements for the following:

- Identification of terminals for grounded conductors
- Use of grounded conductors in premises wiring systems
- Identification of grounded conductors

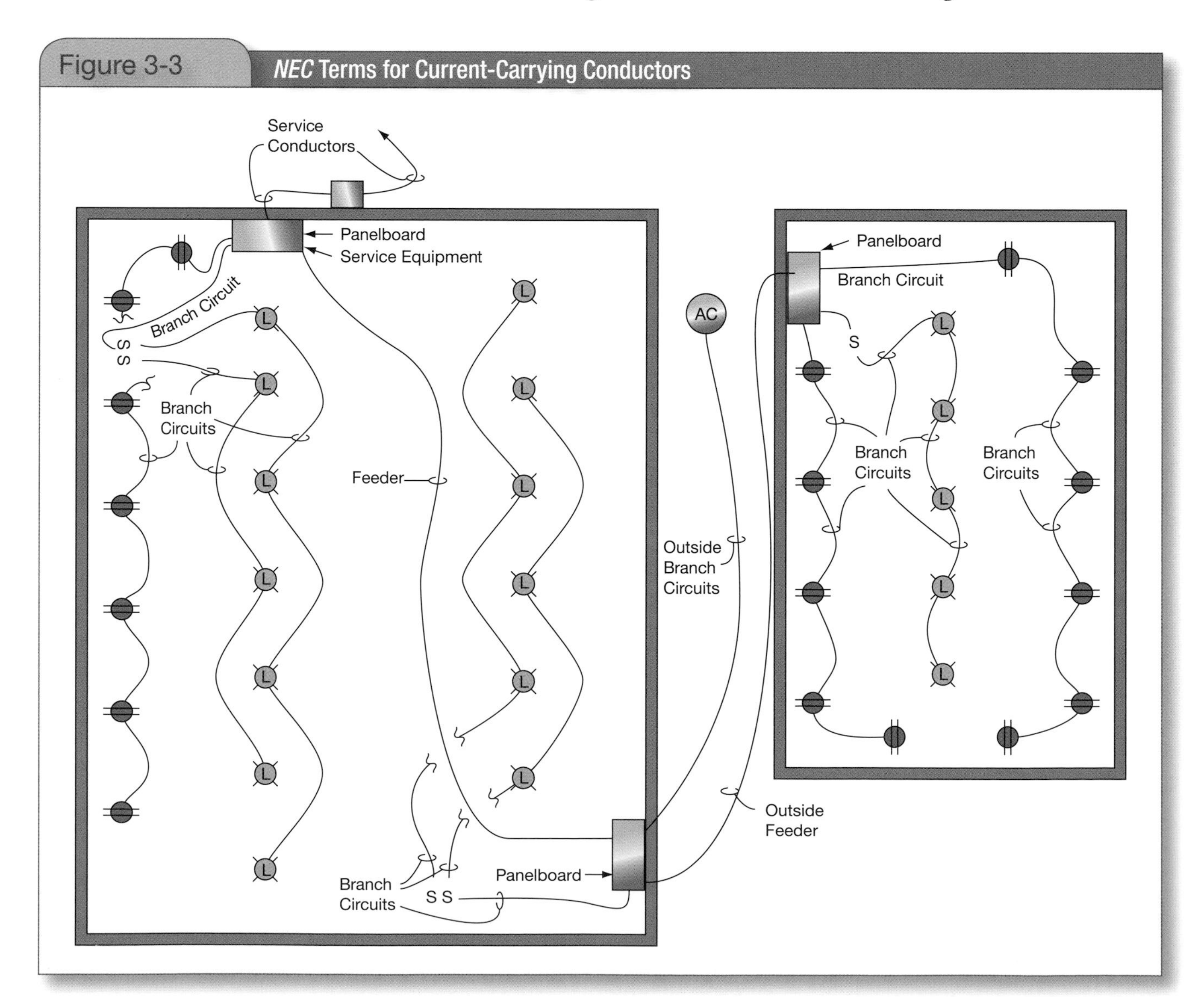

Figure 3-3. The planning requirements in the wiring articles of Chapter 2 focus on the NEC *terms for current-carrying conductors.*

210.19(A)(1) Informative Note No. 4 requires the increase of conductor size for long conductor runs to prevent circuit voltage drop. Long conductor runs increase the circuit's resistance, which in turn reduces the voltage at the equipment being served. The *NEC* allows for up to 3% voltage drop for general branch circuits. A 208-volt source with a 3% voltage drop due to a long conductor run will only provide 202 volts at the equipment being served. Voltage drops greater than 3% would likely damage the equipment being served or cause it to perform poorly. Although it is beyond the scope of this text, a quick/rough calculation for determining the conductor size (CM) to accommodate voltage drop in a long run is:

$$CM = \frac{Ph \times L \times I \times K}{E \times \%VD}$$

Where:

CM: Circular mils (see *NEC* Table 8)

Ph: 1.732 for 3-ph; (for single-phase systems, the multiplier "2" replaces the square root of 3, or 1.732.)

L: The length of the conductor from the source to the load, one way.

I: Current in the circuit

K: 12.9 for copper; 21 for aluminum

E: Nominal voltage

%VD: 0.03 for 3% branch VD

Figure 3-4 **Grounded Conductor**

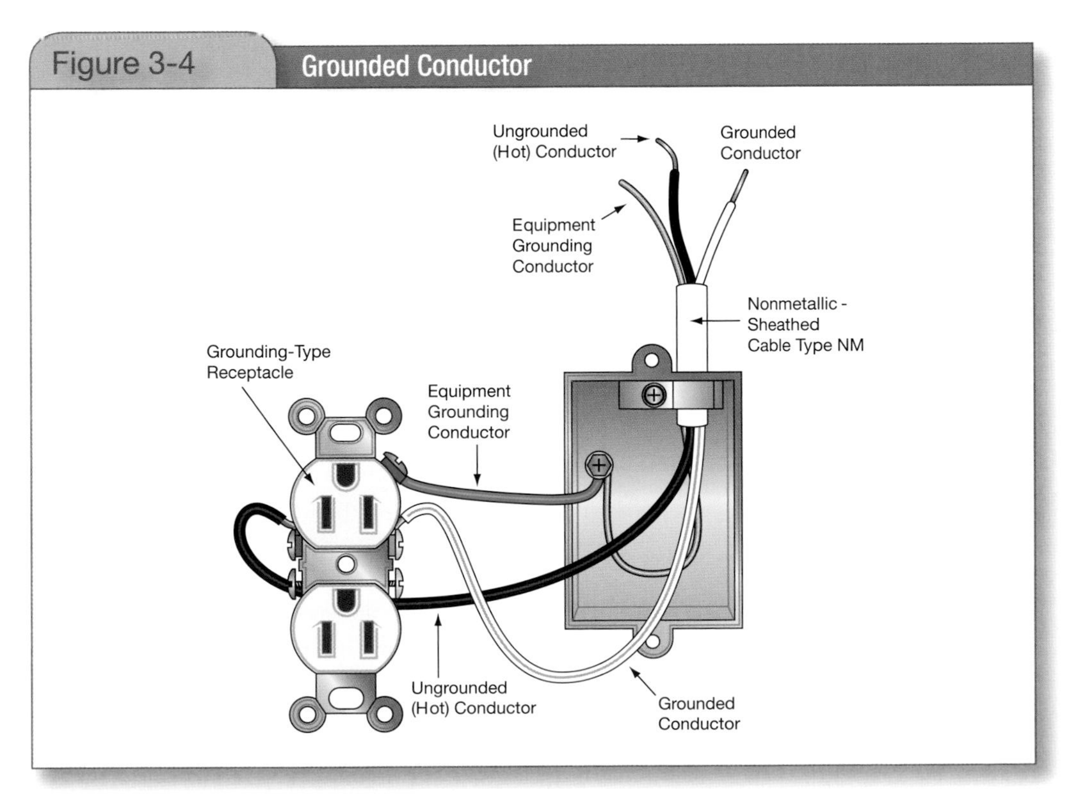

Figure 3-4. The NEC *defines all current-carrying and grounding conductors. The "ungrounded" conductor is commonly called the hot conductor. The "grounded" conductor is commonly called the neutral.*

Article 210 Branch Circuits

The term "branch circuit" is defined in **Article 100**. **See Figure 3-5.**

This basic definition defines the current-carrying conductors that supply current to the load. The different types of branch circuits are defined in **Article 100** to differentiate between the different uses of branch circuits and the specific rules within the *NEC* for each type. **Article 210** provides the general requirements for all branch circuits. These are general requirements only. Requirements for specific branch circuits are found in their respective articles. **Section 210.3** provides a cross-reference table to guide the *Code* user to the correct article for specific-purpose branch circuits.

Article 210 is subdivided into three logical parts as follows:

Part I. General Provisions

- Ratings
- Identification
- Voltage Limitations
- GFCI Requirements
- AFCI Requirements
- Required Branch Circuits

Part II. Branch-Circuit Ratings

- Conductor Ampacity and Size
- Overcurrent Protection
- Outlet Devices
- Permissible Loads
- Common Area Branch Circuits

Part III. Required Outlets

- Dwelling Units
- Guest Rooms
- Show Windows
- Lighting Outlets Required
- HVAC Outlet Required
- Meeting Room Receptacle Outlets

Figure 3-5 Branch Circuits

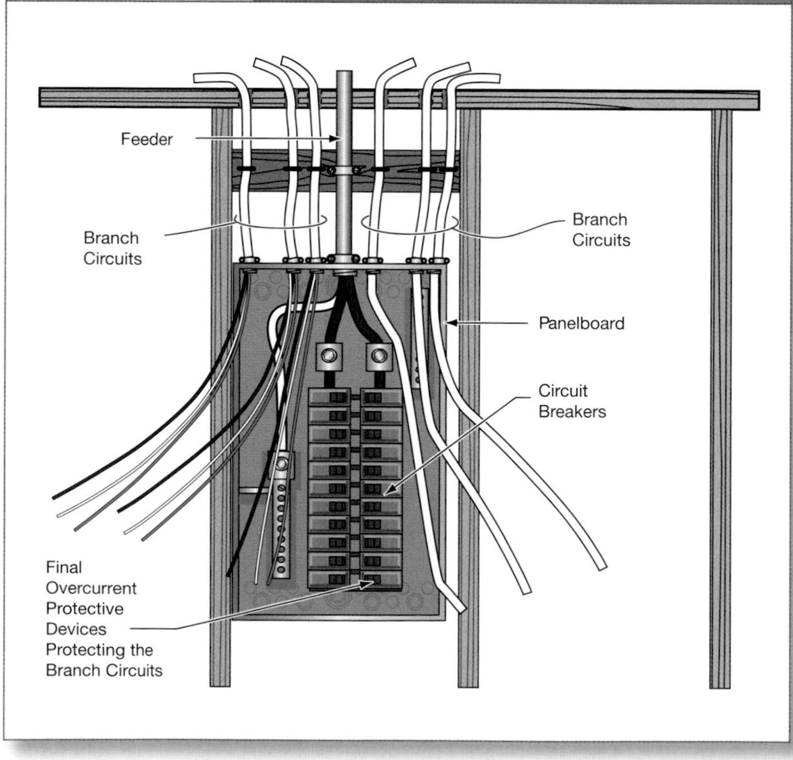

Figure 3-5. The conductors from the final overcurrent device, fuse, or circuit breaker to the receptacles, lighting outlets, hardwired equipment, and all other outlets are branch-circuit conductors.

Article 215 Feeders

Article 100 defines the term "feeder." **See Figure 3-6.**

The three primary types of conductors addressed in the *NEC* are service, feeder, and branch circuit. Service conductors may originate only at a utility-owned and supplied source, and end where disconnecting means and overcurrent protection are provided. Branch circuits begin at the final overcurrent protective device and end at the last outlet or utilization equipment supplied. Feeders are, very simply, all of the conductors in between the service equipment and the final overcurrent protective device.

A hierarchy does not exist for feeders. The term "feeder" is often confused with "subfeeder," a term that does not exist in the *NEC*. Therefore, there is no such thing as a "subfeeder."

Article 215 provides requirements for the following:

- Installation of feeders
- Overcurrent protection for feeders
- Minimum size and ampacity of feeders

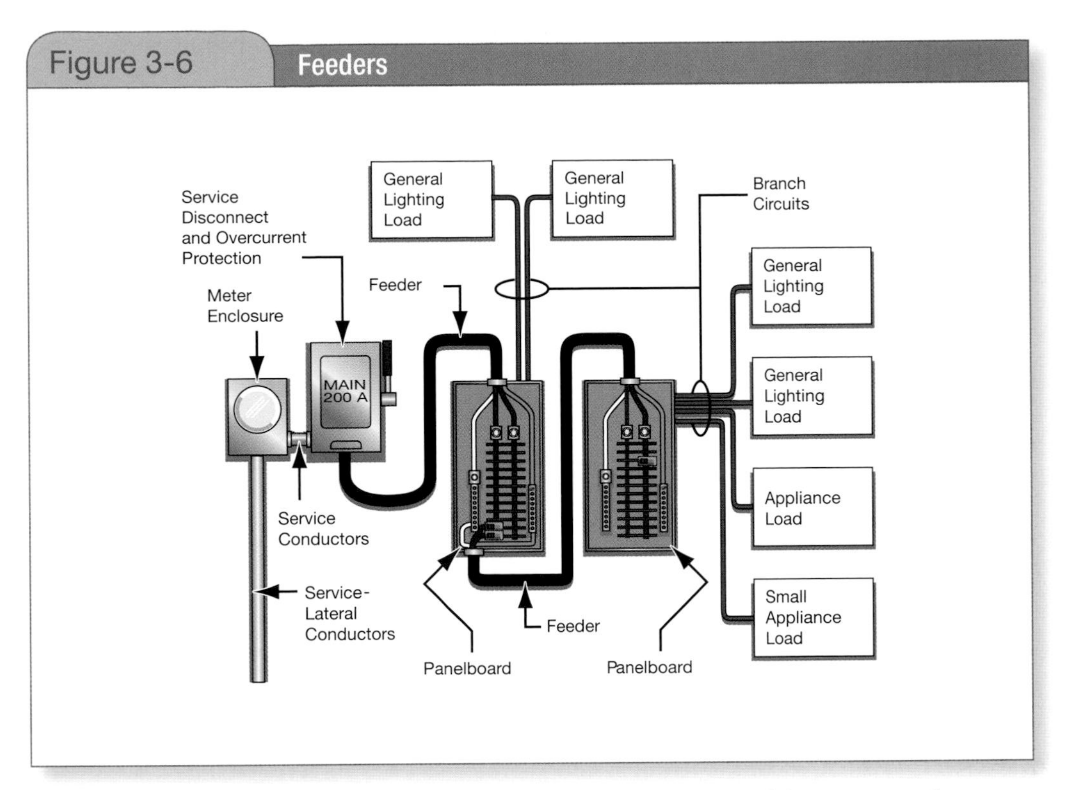

Figure 3-6. All conductors between the final overcurrent protective device and the service or other power supply are feeder conductors.

Article 220 Branch-Circuit, Feeder, and Service Calculations

Article 220 is subdivided into five logical parts as follows:

- General
- Branch-Circuit Load Calculations
- Feeder and Service Load Calculations
- Optional Feeder and Service Load Calculations
- Farm Load Calculation

Article 220 provides requirements for the following:

- Computing branch-circuit loads
- Computing feeder loads
- Computing service loads

Article 225 Outside Branch Circuits and Feeders

While **Article 210** and **Article 215** provide the basic requirements for branch circuits and feeders, **Article 225** provides specific requirements for outside branch circuits and feeders as follows:

- Those run on or between buildings
- Those on structures
- Those on poles on the premises

Article 225 is subdivided into three logical parts as follows:

Part I. General

This part provides general information for outside branch circuits and feeders, including the following:

- Conductor Size and Support
- Outdoor Lighting Equipment
- Overcurrent Protection
- Wiring on Buildings
- Entrance/Exit of Circuits
- Open Conductor Spacing/Support
- Support Over Buildings
- Point of Attachment
- Means of Attachment
- Clearance from Ground
- Protection of Conductors
- Raceways/Cables on Exterior
- Vegetation as Support

Part II. Buildings or Other Structures Supplied by a Feeder(s) or Branch Circuit(s)

This part addresses outside branch circuits and feeders that supply a separate building or structure. Note that a building or structure supplied by an outdoor branch circuit or feeder must comply with rules very similar to those for a service-supplied building or structure.

These basic requirements include the following:

- Number of Supplies
- Location of Disconnecting Means
- Maximum Number of Disconnects
- Grouping of Disconnects
- Accessibility of Disconnects
- Identification of Disconnects
- Disconnect Rating
- Access to Overcurrent Protective Devices

Part III. Over 1000 Volts

Outside branch circuits and feeders operating at over 1000 volts are addressed in **Part III** of **Article 225**. This part provides requirements very similar to those in **Part I** and **Part II**, but modified for systems operating at over 1000 volts.

Article 230 Services

Proper application of **Article 230** requires that the *Code* user be familiar with several **Article 100** definitions related to services. **See Figure 3-7.**

Article 230 is divided into eight logical parts, with each part named for the requirements contained within:

Part I. General

- Number of Services
- Service Conductors Considered as Outside
- Raceway Seals
- Clearance of Service Conductors from Building Openings
- Vegetation as Support

Part II. Overhead Service Conductors

- Insulation or Covering
- Conductor Size and Rating

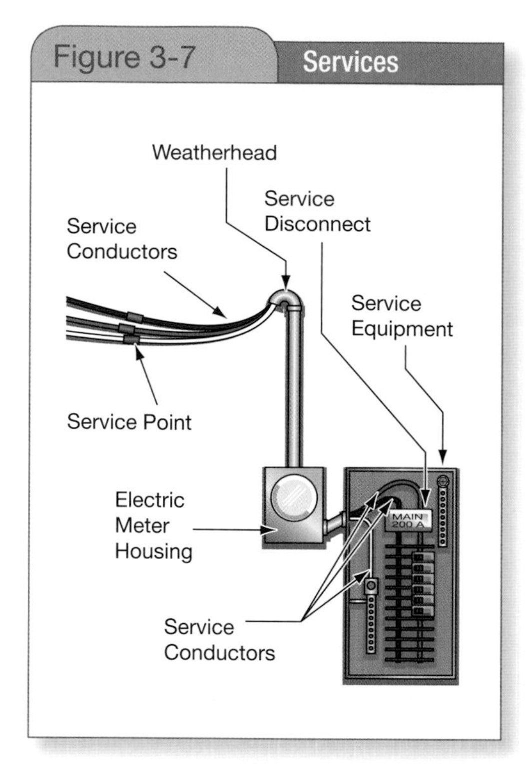

*Figure 3-7. **Article 230** covers all conductors and equipment from the service point to the service disconnect and overcurrent protection.*

- Clearance above Ground and Rooftops
- Point of Attachment to Building
- Means of Attachment
- Service Masts as Supports
- Supports over a Building

Part III. Underground Service Conductors

- Insulation
- Conductor Size and Rating
- Splices
- Protection

Part IV. Service-Entrance Conductors

- Number of Sets Served
- Insulation
- Protection
- Conductor Size and Rating
- Permitted Wiring Methods
- Splices
- Cable Trays
- Supports
- Raceways Arranged to Drain

- Overhead Connections Service Head/ Gooseneck, Drip-loops
- 4-wire Delta Configured Systems

Part V. Service Equipment – General
- Enclosure Requirements
- Marking of Equipment, Suitable as "Service Equipment"

Part VI. Service Equipment – Disconnecting Means
- Disconnect Requirement
- Location of Disconnect Marking
- Accessibility of Disconnect
- Maximum Number of Disconnects
- Grouping of Disconnects
- Operation of Disconnects
- Indicating (Identifies Open or Closed)
- Rating of Disconnect (amps)
- Equipment Permitted on the Line Side (Upstream) of the Service Disconnect

Part VII. Service Equipment – Overcurrent Protection
- Where Required
- Location
- Locking of Service OCPD
- Ground Fault Protection of Equipment (GFPE)

Part VIII. Services Exceeding 1000 Volts, Nominal
- General Requirements
- Isolating Switches
- Disconnects
- Protection
- Surge Arresters
- Metal Enclosed Switchgear
- Services over 35,000 Volts

PROTECTION

In the planning stages of any electrical installation, protection must be incorporated in the electrical drawings as part of the design. The drawings provide the installer with the necessary information to build an installation that is properly protected. All electrical installations must provide overcurrent protection for all current-carrying conductors, which must be protected in accordance with the conductor ampacity and conditions of use in accordance with **Article 240 Overcurrent Protection**.

Grounding of electrical systems is designed to limit the voltage imposed by lightning, line surges, or unintentional contact with higher voltage lines and to stabilize the voltage to earth during normal operation. Normally non–current-carrying metal parts of electrical systems are connected together (bonded) and are connected to the electrical supply source in a manner that creates an effective ground-fault current path. Systems and circuits are required to be grounded in accordance with **Article 250 Grounding and Bonding**. Protection from surge voltages many times attributed to lightning is provided in electrical systems through the use of **Article 280 Surge Arrester, Over 1000 Volts**. Protection from voltage surges at levels closer to nominal voltage is provided for by the application of **Article 285 Surge-Protective Devices (SPDs), 1000 Volts or Less**.

Code-compliant electrical installations must provide adequate protection of persons and property in accordance with the *NEC*. The required protection for electrical installations that must be planned include the following key words/clues:
- Overcurrent protection, fuses, circuit breakers
- Grounding and bonding of systems and circuits
- Grounding and bonding raceways, equipment, or enclosures
- Surge arresters, over 1000 Volts
- Surge-protective devices (SPDs), 1000 Volts or less

FACT

Surge Protection Devices (SPDs) are identified by where they are located in the circuit. SPDs are devices (**Article 100**) that protect electronic equipment from harmful transient currents generated by lightning and large equipment starting and stopping. See **Article 100** for the definition of SPDs.

Articles dedicated to protection can also be found in Chapter 2. **See Figure 3-8**.

Article 240 Overcurrent Protection

Article 100 defines the term "overcurrent."

This basic definition must be understood to properly apply the provisions of **Article 240**. All conductors and equipment are rated for the maximum amount of current (amps) that it can handle without suffering damage. An overcurrent occurs when one of three incidents happens: an overload, a short circuit, or a ground fault.

Overloads occur when a conductor or equipment is subjected to a current exceeding its ampere rating. Overload current stays on the normal circuit path, which, if allowed to continue for too long, would cause damage or dangerous overheating. A fault, such as a short circuit or ground fault, is not an overload. A conductor rated at 20 amps with 22 amps of current flowing would be experiencing an overload. An overload never leaves the circuit path. Current continues to flow from the source through the circuit conductors and load back to the source.

A short circuit occurs when the current leaves the normal circuit path and takes a shortcut back to the source. Short circuits occur when current-carrying conductors make contact with each other, thus creating a shortcut for current flow. Any combination of two or more circuit conductors (current-carrying) in contact results in a short circuit. Current-carrying conductors include all ungrounded (hot) and grounded (in most cases, neutral) conductors.

A ground fault is a form of short circuit. The current takes a shortcut on a grounded component, such as a raceway, enclosure, or building steel, or an equipment grounding or bonding conductor.

The scope of **Article 240 Overcurrent Protection** is to provide general requirements for overcurrent protection and overcurrent protective devices. Overcurrent protective devices (OCPDs) include, but are not limited to, fuses and circuit breakers. **See Figure 3-9**. These devices are designed to provide protection from overloads, short circuits, and ground faults.

Article 240 consists of nine logical parts as follows:

Part I. General

- Definitions
- Cross-Reference Other Articles
- Protection of Conductors
- Protection of Flexible Cords/Cables
- Standard OCPD amp Ratings
- Fuses or Circuit Breakers in Parallel
- Supplementary Overcurrent Protection
- Thermal Devices
- Electrical System Coordination
- Ground-Fault Protection of Equipment

Part II. Location

- Ungrounded Conductors
- Location of Overcurrent Protection in Circuit, Tap Rules
- Grounded Conductors
- Location/Accessibility of OCPDs

Part III. Enclosures

- General Protection and Operation
- Damp or Wet Locations
- Vertical Position

Figure 3-8 — Chapter 2 (Wiring and) Protection

***NEC* Title:**	**Wiring and Protection**
***Codeology* Title:**	**Plan**
Chapter Scope:	**Information and Rules on Wiring and Protection of Electrical Installations**

PROTECTION

Article	Article Title
240	Overcurrent Protection
250	Grounding and Bonding
280	Surge Arresters, Over 1000 Volts
285	Surge-Protective Devices (SPDs), 1000 Volts or Less

*Figure 3-8. **Articles 240, 250, 280,** and **285** cover the protection section of Chapter 2.*

Figure 3-9 OCPDs

Figure 3-9. Fuses and circuit breakers provide overcurrent protection for conductors and equipment.

For additional information, visit qr.njatcdb.org Item #1067

Part IV. Disconnecting and Guarding
- Disconnects for Fuses
- Arcing or Suddenly Moving Parts

Part V. Plug Fuses, Fuseholders, and Adapters
- General Application
- Edison Base Fuses
- Type "S" Fuses

Part VI. Cartridge Fuses and Fuseholders
- General Application
- Classification

Part VII. Circuit Breakers
- Method of Operation
- Indicating
- Nontamperable
- Marking
- Applications
- Series Ratings

Part VIII. Supervised Industrial Installations
This part addresses only those portions of a building or structure that meet the conditions of a supervised industrial installation, as defined in **240.2**.

Part IX. Overcurrent Protection Over 1000 Volts, Nominal
This part is limited only to feeders and branch circuits operating at over 1000 volts nominal.

Article 250 Grounding and Bonding

The scope of **Article 250** is provided in the first section of the article. **See Figure 3-10.**

250.4 General Requirements for Grounding and Bonding

(A) Grounded Systems
(1) Electrical System Grounding
(2) Grounding of Electrical Equipment
(3) Bonding of Electrical Equipment
(4) Bonding of Electrically Conductive Materials and Other Equipment
(5) Effective Ground-Fault Current Path

(B) Ungrounded Systems
(1) Grounding Electrical Equipment
(2) Bonding of Electrical Equipment
(3) Bonding of Electrically Conductive Materials and Other Equipment
(4) Path for Fault Current

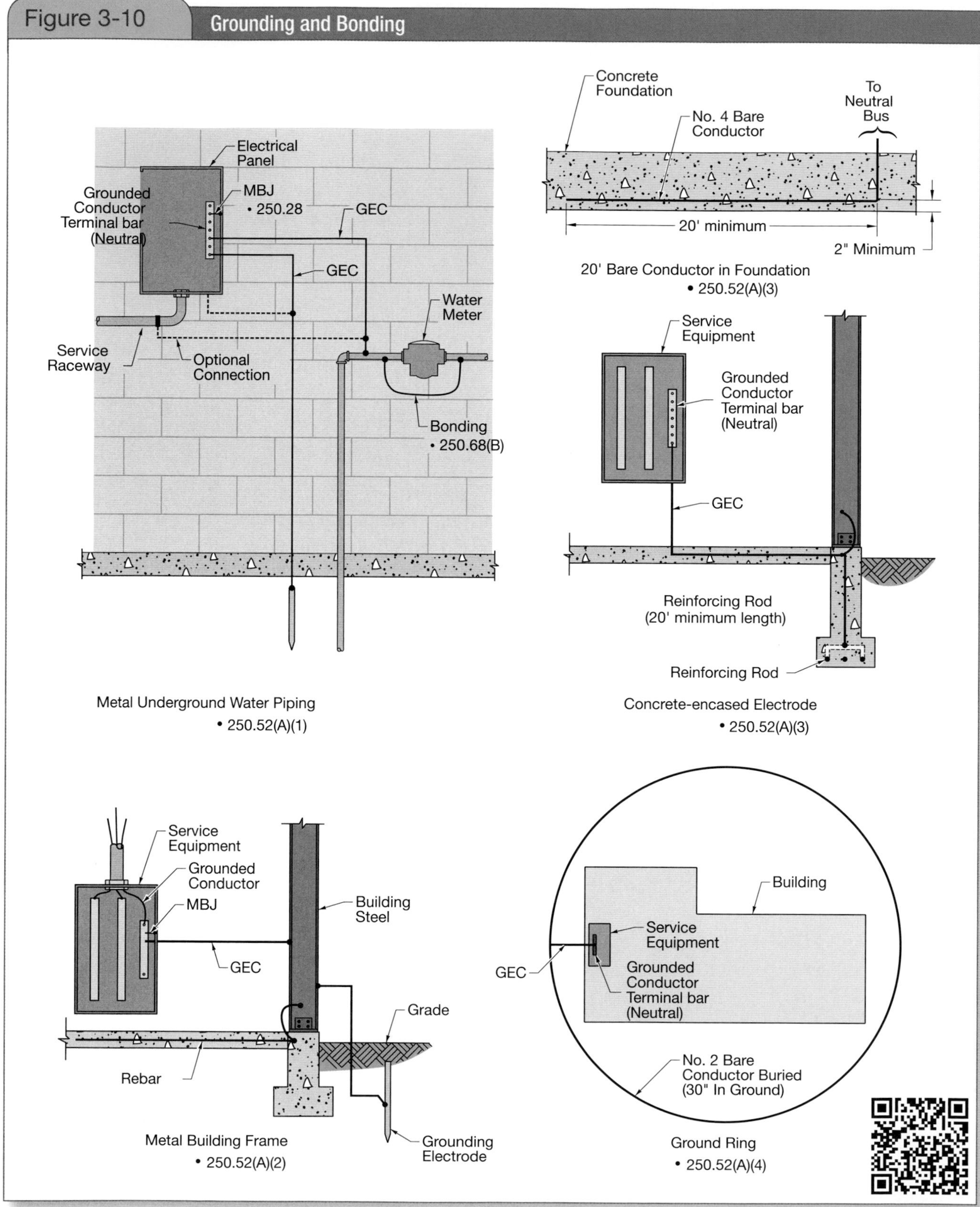

For additional information, visit qr.njatcdb.org Item #1068

Figure 3-10. Article 250 covers protection requirements through provisions for grounding and bonding of electrical systems and installations.

Article 250 is subdivided into ten logical parts as follows:

Part I. General
- Definitions
- Cross-Reference Other Articles
- Reasons for Grounding and Bonding Connections
- Protection of Clamps/Fittings
- Clean Surfaces

Part II. System Grounding
- Systems Requiring Grounding
- Systems not Requiring Grounding
- Circuits not Permitted to Be Grounded
- Grounding AC Services
- Conductor to be Grounded
- Main Bonding Jumpers
- Separately Derived Systems
- Two or More Buildings with Common Service
- Generators: Portable and Permanent
- High-Impedance Systems

Part III. Grounding Electrode System and Grounding Electrode Conductor
- Outline of the Grounding Electrode System
- Permitted Electrodes
- Installation of Grounding Electrode System
- Common Electrodes
- Supplementary Electrodes
- Resistance of Electrodes
- Air Terminals
- Size of Grounding Electrode Conductor
- Connection of Grounding Electrode Conductors

Part IV. Enclosure, Raceway, and Service Cable Connections
- Service Raceways and Enclosures
- Underground Service Cable and Conduit
- Other Enclosures and Raceways

Part V. Bonding
- Services
- Other Systems
- Other Enclosures
- Over 250 Volts
- Loosely Jointed Raceways
- Hazardous Locations
- Equipment Bonding Jumpers, Supply and Load Side
- Piping Systems and Exposed Structural Steel
- Lightning Protection Systems

Part VI. Equipment Grounding and Equipment Grounding Conductors
- Equipment Fastened in Place
- Cord-and-Plug-Connected Equipment
- Nonelectric Equipment
- Types of Equipment Grounding Conductors
- Identification of Equipment Grounding Conductors and Device Terminals
- Installation
- Size of Equipment Grounding Conductors

Part VII. Methods of Equipment Grounding Connections
- Short Sections of Raceway
- Ranges/Clothes Dryer Frames
- Equipment Fastened in Place
- Cord and Plug Connected Equipment
- Use of Grounded Conductor
- Receptacle Grounding Attachment to Boxes
- Continuity of Equipment Grounding Conductor

Part VIII. Direct-Current Systems
- Circuits and Systems to Be Grounded
- Point of Connection
- Size of Grounding Electrode Conductor
- Bonding Jumpers

Part IX. Instruments, Meters, and Relays
- Transformer Circuits and Cases
- Cases of Equipment at Over 1000 Volts
- Grounding Conductors

Part X.Grounding of Systems and Circuits over 1000 Volts
- General Requirements
- Derived Neutral Systems

- Grounding
- Solidly Grounded Neutral Systems
- Impedance Grounded Neutral Systems
- Portable or Mobile Equipment

Article 280 Surge Arresters, Over 1000 Volts

Surge arrester is defined in **Article 100. See Figure 3-11.**

Article 280 consists of three parts as follows:

Part I. General

- Uses Not Permitted
- Listing
- Number of Surge Arresters Required
- Selection

Part II. Installation

- Location of Surge Arresters
- Routing Surge Arrester Connections

Part III. Connecting Surge Arresters

- Grounding

Article 285 Surge-Protective Devices (SPDs), 1000 Volts or Less

Article 100 defines a *surge-protective device* (SPD). **See Figure 3-12.**

Article 285 consists of three logical parts as follows:

Part I. General

- Uses Not Permitted
- Number Required
- Listing Requirements
- Short Circuit Ratings

Part II. Installation

- Location
- Routing of Connections

Part III. Connecting SPDs

- Connection of SPDs
- Grounding

Figure 3-11 Surge Arresters

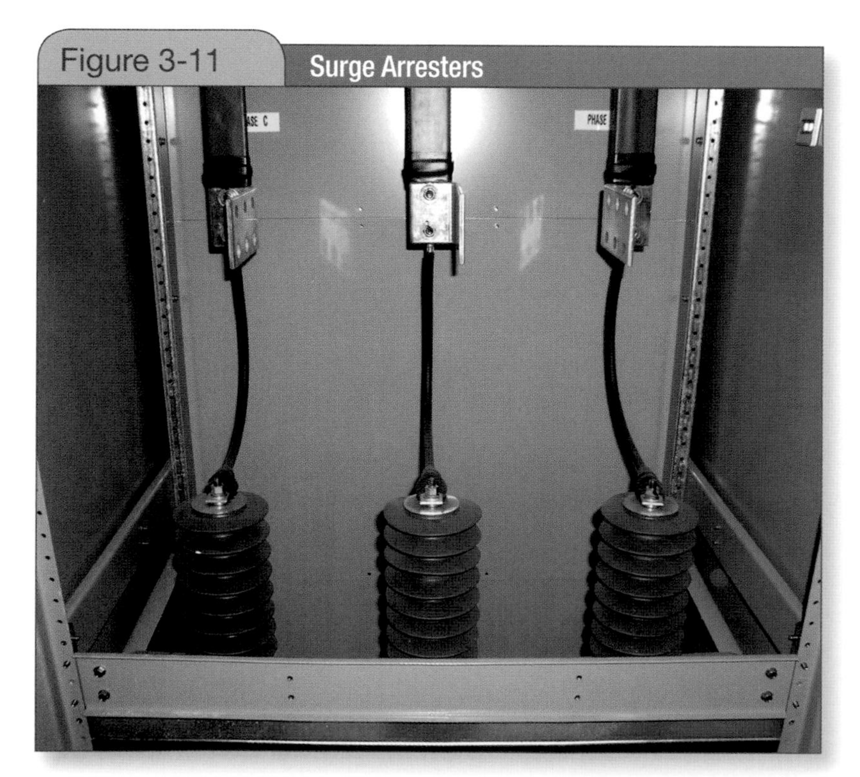

Figure 3-11. Surge arresters limit surge voltages by discharging or bypassing dangerous surge voltages to ground.

Figure 3-12 SPDs 1000 V or Less

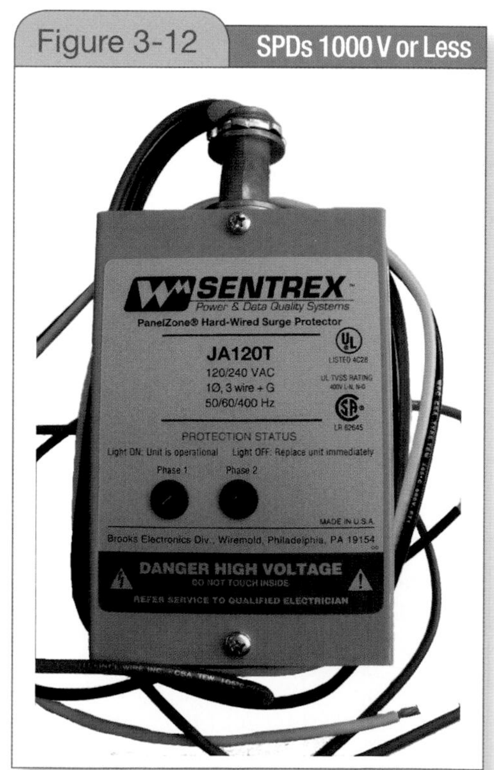

Figure 3-12. Surge-protective devices (SPDs) provide protection from an overvoltage at levels much closer to the operating voltage than surge arresters.

Figure 3-13 Key Words and Clues

WIRING	PROTECTION
• Branch circuits, all types GFCI	• Overcurrent protection, fuses, circuit breakers
• Branch circuits, indoor/outdoor	• Grounding systems and circuits
• Required outlets AFCI	• Grounding and bonding raceways, equipment or enclosures
• Feeders, indoor/outdoor	• Surge arresters
• Service/s service equipment	• Surge-protective devices
• Calculations, computed load	• Equipment grounding conductors
• Grounded conductors, neutral	• Location of overcurrent protective devices

Figure 3-13. Key words and clues regarding wiring and protection will lead the Code *user to review Chapter 2 of the* NEC.

KEY WORDS AND CLUES FOR CHAPTER 2, "PLAN"

A list of key words and clues for Chapter 2 are included for reference. **See Figure 3-13.** Chapter 2 topics are many of the most common *Code* requirements in daily installations. As previously discussed, grounding and bonding apply to practically every electrical installation. It is paramount that electricians have a solid understanding of Chapter 2 and how to locate sections based upon key words and clues.

Refer to Chapter 2 when the question or need within the *NEC* deals with any of the following:

- Wiring, *NEC* current-carrying conductor names, branch circuit, feeder, service or tap conductor
- Grounded Conductors
- Ground-Fault Circuit Interrupters (GFCIs)
- Arc-Fault Circuit Interrupters (AFCIs)
- Required Outlets
- Calculations, Computed Loads
- Outside Branch Circuits and Feeders
- Protection, Overcurrent, Grounding, Surge Arresters, and Surge-Protective Devices
- Fuses and Circuit Breakers
- Grounding and Bonding Conductors
- Grounding Electrodes

Summary

The *Codeology* method is used to locate information in the *NEC* quickly and accurately. Using clues and key words in a question and then applying the generic terms for the *NEC* Chapters is the starting point. Knowing topics in the *Code* and chapter/article titles is a key point for success. Using the *Codeology* method is a systematic, step-by-step approach that involves first finding the key words to a question, going to the Table of Contents and finding the correct part, and then reading the bold type in the chapter, article, and part to find the answer to an inquiry.

Summary

Chapter 2, in accordance with **Section 90.3**, applies generally to all electrical installations. Using the *Codeology* method, this chapter has been called the "Plan" chapter, due to its wide scope of coverage. **Chapter 2 Wiring and Protection** has been described as information and rules on wiring and protection of electrical installations.

Chapter 2 covers the entire electrical system from the service point (connection to the utility) to the last receptacle or other outlet in the electrical system. Chapter 2, in accordance with its scope of "Wiring," provides detailed requirements for all current-carrying conductors. To provide clarity, the *NEC* names and defines these conductors. Names include service, feeder, branch-circuit, and tap conductors.

In accordance with its scope of "Protection," Chapter 2 provides detailed requirements to protect the entire electrical system. These protection requirements include overcurrent, grounding, bonding, surge arresters, and surge-protective devices.

Chapter 2, along with Chapters 1, 3, and 4, builds the foundation or backbone of all electrical installations. Chapter 2 will apply generally in all electrical installations.

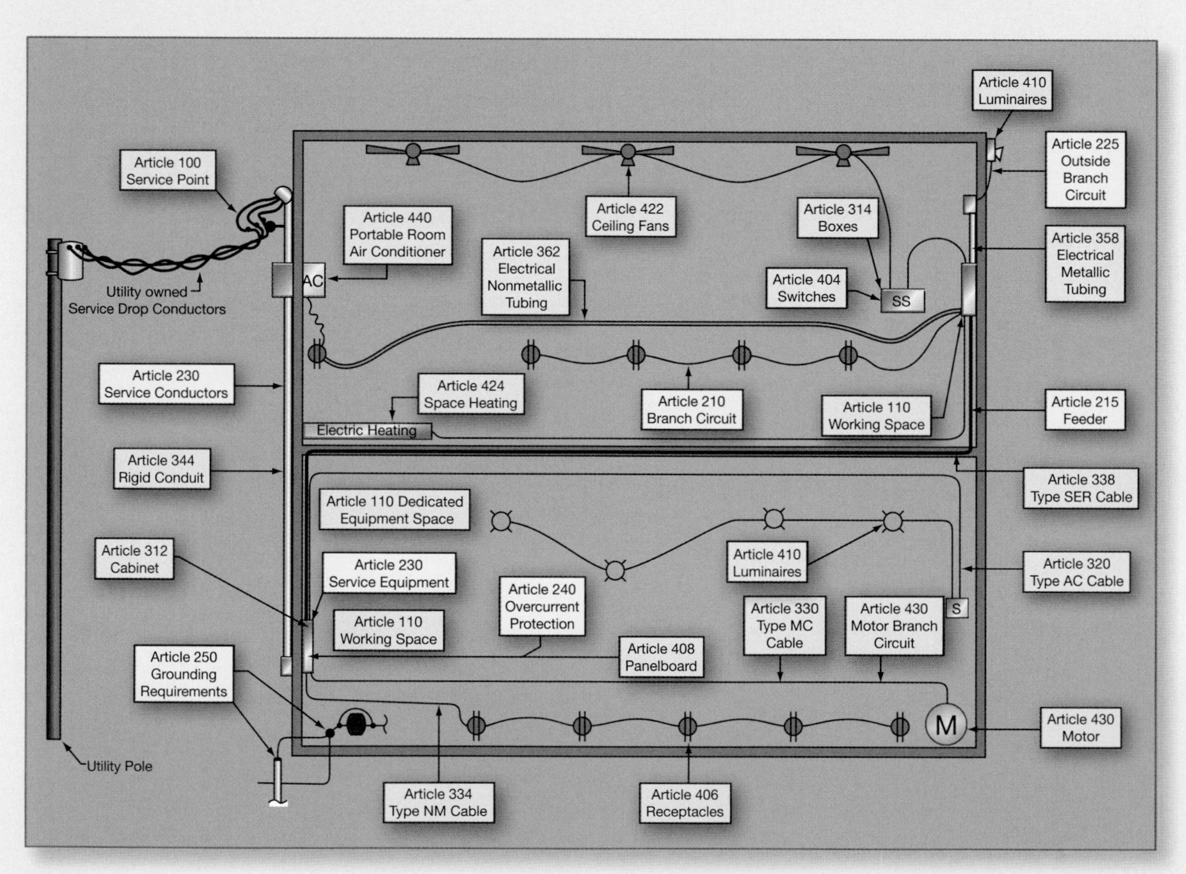

Chapter 2 will apply generally in all electrical installations.

Review Questions

1. **Why is the *Codeology* method used for finding information or answers to questions?**
   a. It is a method of checks and balances to provide answers to questions.
   b. It is an organized method of using the structure outline format of the *NEC*.
   c. It is the easiest method to find research information available to the industry.
   d. It uses the index as a foundation to find *Code* sections.
2. **Chapters 2 and 3 of the *NEC* may be considered the "heart" or "backbone" of the *Code* installations process.**
   a. True
   b. False
3. **Chapter 2 of the *NEC* applies __?__.**
   a. generally to all of the chapters in the *NEC*
   b. generally to Chapters 1, 2, 3, and 4
   c. only to the information contained in Chapter 2
   d. none of the above
4. **Chapter 2 is sub-divided into __?__.**
   a. 10 articles
   b. 11 articles
   c. 44 articles
   d. none of the above
5. **The scope of Chapter 2 of the *NEC* consists of __?__ articles covering wiring and __?__ articles covering protection.**
   a. 4 / 40
   b. 5 / 6
   c. 6 / 4
   d. 40 / 4
6. **The part and article of Chapter 2 that addresses an outdoor branch circuit serving a second structure is __?__.**
   a. **Article 225 Part I**
   b. **Article 225 Part II**
   c. **Article 225 Part IV**
   d. **Article 225 Part V**
7. **Article 240 is divided into how many parts?**
   a. 8
   b. 9
   c. 10
   d. 45
8. **Part V of Article 240 applies only to __?__.**
   a. adapters
   b. fuseholders
   c. plug fuses
   d. all of the above
9. **Which part of Article 250 would contain requirements for grounding electrode conductors?**
   a. **Part I**
   b. **Part III**
   c. **Part VI**
   d. **Part X**
10. **Which of the following articles is not one of the "wiring" articles of Chapter 2?**
    a. **210**
    b. **225**
    c. **240**
    d. **250**
11. **Which one of the following articles belongs in the "protection" section of Chapter 2?**
    a. **200**
    b. **210**
    c. **220**
    d. **285**

The *NEC* "Build" Chapter (Wiring Methods and Materials)

Introduction

The *Codeology* title for Chapter 3 is "Build." Electrical installations can only occur after they have been properly planned. Chapter 3 provides the rules and information necessary for all of the wiring methods and materials used to distribute electrical energy. The *Codeology* generic term used for Chapter 3 is "Build." Only the wiring methods and materials allowed by *Code,* such as conductor types, boxes, enclosures, cable assemblies, raceways, etc., are contained in this chapter. Electrical equipment, such as switches, receptacles, panelboards, switchboards, motors, lighting, and appliances, are not contained within the scope of this chapter and will be discussed in the "Use" chapter. These types of equipment, which provide for control, transformation, and utilization, are addressed in Chapter 4 of the *NEC*.

Objectives

- » Associate the *Codeology* title for *NEC* Chapter 3 as "Build."
- » Recite from memory the recommended articles and their titles that form *NEC* Chapter 3.
- » Recognize Chapter 3 numbering as the 300-series.
- » Understand the building and hands-on type of information and requirements for wiring methods and wiring materials contained in Chapter 3.
- » Identify Chapter 3 numbering as the 300-series.
- » Recognize, recall, and apply the articles contained in Chapter 3.
- » Use the *Codeology* method to find information in *NEC* Chapter 3.

Chapter 4

Table of Contents

NEC CHAPTER 3, "BUILD"

Chapter 3, the 300-series, contains 46 articles. These articles address the methods and materials for the distribution of electrical energy. In essence, Chapter 3 covers all methods and materials necessary to connect electrical energy from the power source to the electrical equipment. Because this chapter covers electrical work where wire and materials are used, it is sometimes referred to as a mechanical work or "installation by the electrician" chapter. It can also be called the mechanical distribution means by which electrical energy is delivered through circuit conductors (a common term used throughout the *Code* meaning "current-carrying conductors"), or simply referred to as the "pipe and wire" chapter.

The wiring methods and materials covered in the "Build" chapter include the following:

- **300** – General installation rules
- **310** – Conductors
- **312-314** – Enclosures, cabinets, boxes
- **320-340** – Cable assemblies
- **342-362** – Circular raceways, conduits
- **366-370** – Auxiliary Gutters/Busways/Cablebus
- **372-393** – Other raceways or supporting systems
- **394-399** – Open wiring

The materials used to distribute electrical energy throughout buildings and

Commit the following articles in Chapter 3 to memory to better facilitate use of the *NEC*.

- **300** – General Requirements for Wiring Methods and Materials.
- **310** – Conductors for General Wiring
- **314** – Outlet, Device, Pull and Junction Boxes; Conduit Bodies; Fittings; and Handhole Enclosures
- **320** – Armored Cable: Type AC
- **330** – Metal-Clad Cable: Type MC
- **334** – Nonmetallic-Sheathed Cable: Type NM, NMC, NMS
- **338** – Service-Entrance Cable: Type SE and USE
- **342** – Intermediate Metal Conduit: Type IMC
- **344** – Rigid Metal Conduit: Type RMC
- **348** – Flexible Metal Conduit: Type FMC
- **350** – Liquid Tight Flexible Metal Conduit: Type LFMC
- **352** – Rigid Polyvinyl Chloride Conduit: Type PVC
- **358** – Electrical Metallic Tubing: Type EMT
- **368** – Busways
- **392** – Cable Tray

Figure 4-1a Distribution of Electrical Energy

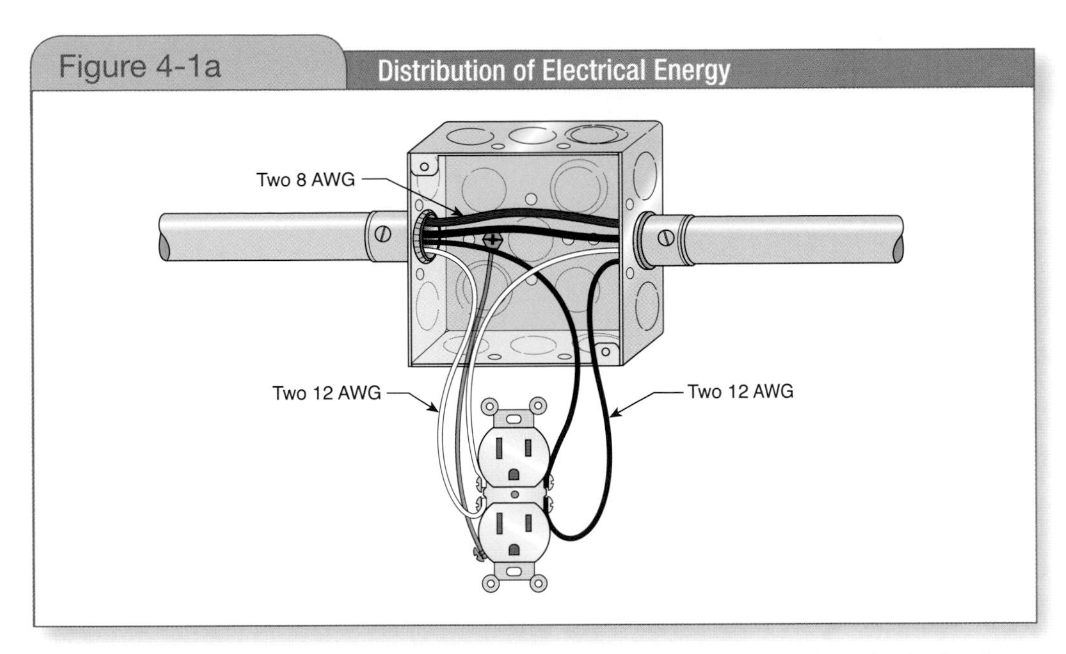

*Figure 4-1a. Junction boxes (wiring materials), as listed in **Article 314**, are commonly used in the distribution of the conduit system that provides electrical service to outlets and equipment.*

Figure 4-1b Conductors

Figure 4-1b. Conductors (wiring methods) of various sizes and insulation levels are often installed together.

Figure 4-1c EMT Conduit

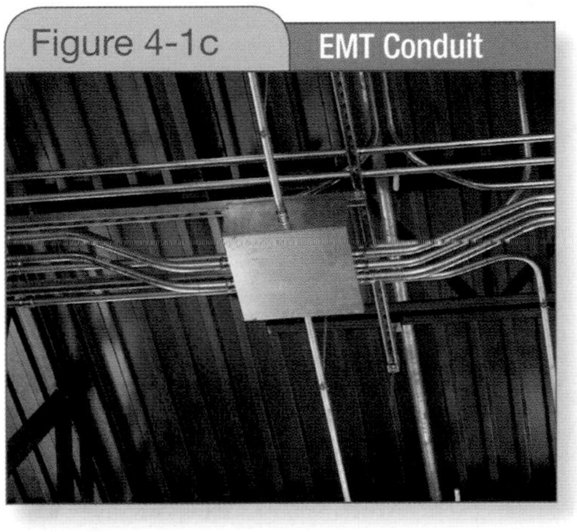

Figure 4-1c. EMT conduit (Wiring Methods) is often used to distribute power in commercial installations.

structures are broken down to *wiring methods* and *wiring materials*. These two topics describe the basic roots of Chapter 3. *Wiring methods* are the materials that carry the current (for example, conductors) along the mechanical means to carry and route these conductors (for example, conduits). *Wiring materials* facilitate the electrical installation by utilizing items such as junction boxes, hangers, and other support items. Chapter 3 of the *NEC* covers various types of conductors and cables, raceways (for example, conduit), and junction boxes that are used in electrical installations. **See Figure 4-1.**

The requirements and information in *NEC* Chapter 3 are logically separated into 13 different categories. Practically every electrical installation requires current-carrying conductors from the power source to the final equipment. These conductors must be protected and

Figure 4-2 **Categories of the *NEC***

***NEC* Title:**	**Wiring Methods and Materials**
***Codeology* Title:**	**Build**
Chapter Scope:	**Information and Rules on Wiring Methods and Materials for Use in Electrical Installations**
Category 1	General Information for All Wiring Methods and Materials
Category 2	Conductors
Category 3	Cabinets, Boxes, Fittings, and Meter Socket/Handhole Enclosures
Category 4	Cable Assemblies
Category 5	Raceways, Circular Metal Conduit
Category 6	Raceways, Circular Nonmetallic Conduit
Category 7	Raceways, Circular Metallic Tubing
Category 8	Raceways, Circular Nonmetallic Tubing
Category 9	Factory-Assembled Power Distribution Systems
Category 10	Raceways Other than Circular
Category 11	Surface-Mounted Nonmetallic Branch Circuit Extensions
Category 12	Support Systems for Cables/Raceways
Category 13	Open-Type Wiring Methods

Figure 4-2. *Articles of Chapter 3 describe wiring methods and wiring materials for electrical installations.*

routed in a physically protective means, typically by a conduit or cable covering. **See Figure 4-2.**

WIRING METHODS

Wiring methods are the materials used to conduct current throughout an electrical installation. A wiring method is any combination of conductors, such as copper wire, with a protective means or layer of installation, such as THHN. Very often the conductors are grouped together under one protective means or layer and referred to as a cable assembly (or simply called a cable). For further clarification, a conductor is a single wire, whereas a cable is composed of multiple wires under one protective means or layer.

Cable assemblies are common wiring methods. **See Figure 4-3.** For example, type AC cable consists of insulated conductors, wrapped in paper, with a flexible corrugated-metal outer sheath (or armor) of aluminum or steel that includes an internal bonding strip of copper or aluminum in intimate contact with the armor

Figure 4-3a **Wiring Methods**

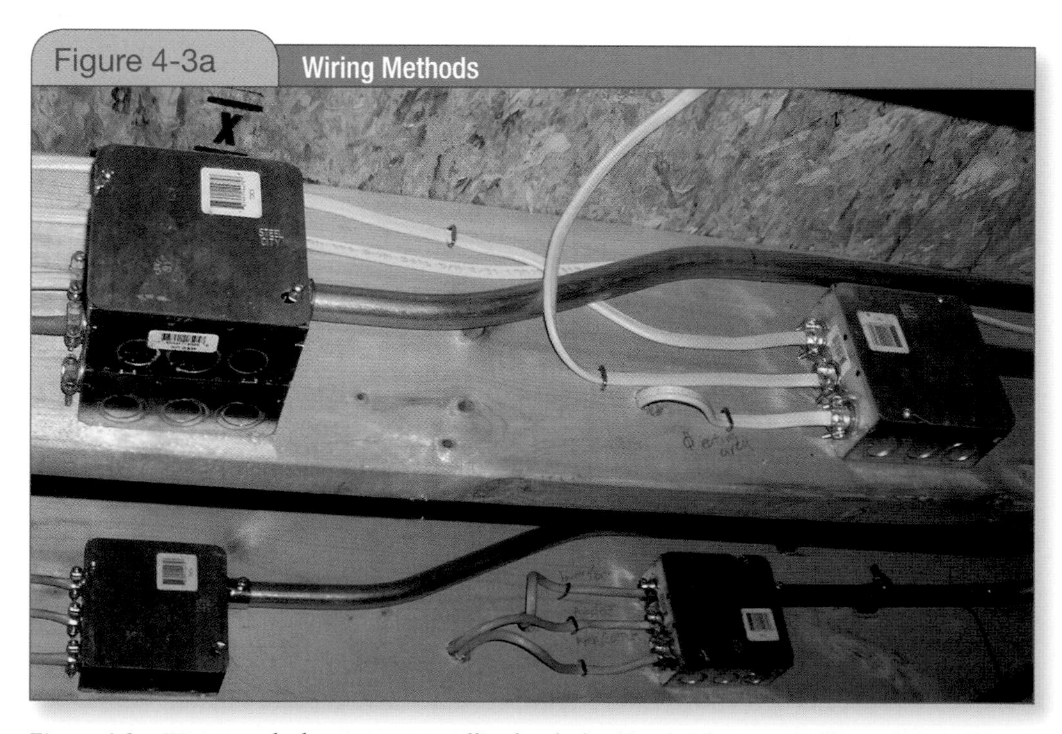

Figure 4-3a. *Wiring methods using nonmetallic-sheathed cables (NM) are typically used in residential installations.*

Figure 4-3b Wiring Methods

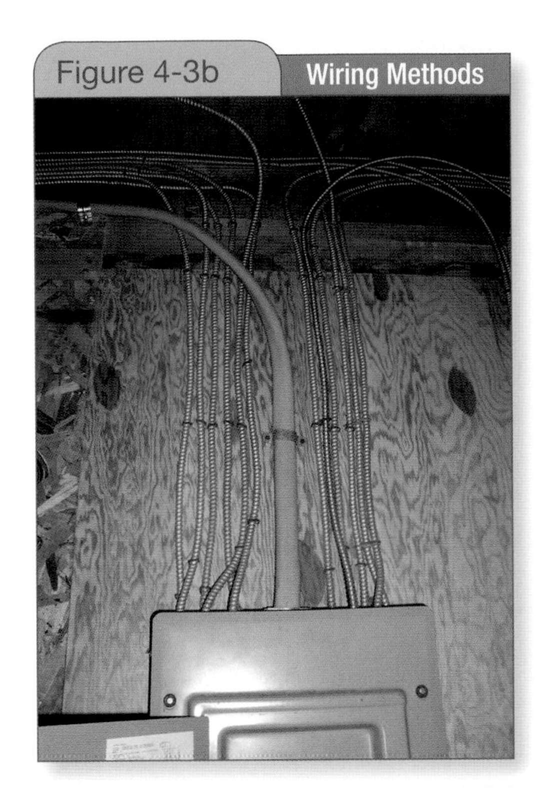

Figure 4-3b. Wiring methods using metal-clad cables (MC) are typically used in commercial installations.

for its entire length. Eleven types of cable assemblies are covered in Chapter 3 and are listed in alphabetical order from **Articles 320** to **340**.

Raceways, unlike cable assemblies, do not come from the manufacturer prewired and must have conductors installed after their installation. Raceways are also considered wiring methods. For example, electrical metallic tubing, type EMT, with insulated type THHN conductors installed, is a wiring method. Chapter 3 covers 25 types of raceways in five different categories. **See Figure 4-4.**

The decision of which of the five categories of raceway to use in an installation is based upon several factors, such as economics, required physical protections, environmental location, building construction, and even AHJ requirements. In some cases, it may be based upon what the customer desires, as long as it is permissible under the *NEC*. The installation practices are certainly different between metal and PVC raceways. More than one type of raceway is often used on an electrical project. **See Figure 4-5.**

For example, a branch circuit feeding an electric sign in the front yard of

Figure 4-4 Raceway Categories

Categories of Raceways	Example
Raceways, Circular Metal Conduit	Rigid Metallic Conduit (RMC)
Raceways, Circular Nonmetallic Conduit	Polyvinyl Chloride (PVC) Tubing
Raceways, Circular Metallic Tubing	Electrical Metallic Tubing (EMT)
Raceways, Circular Nonmetallic Tubing	Electrical Nonmetallic Tubing (ENT)
Raceways Other than Circular	Gutters; Metallic Wireways

Figure 4-4. Various means of raceways are used in electrical installations.

Figure 4-5a RMC (Article 344)

Figure 4-5a. Rigid metal conduit provides maximum physical protection.

Figure 4-5b PVC (Article 352)

Figure 4-5b. Circular nonmetallic conduit is mainly utilized underground and outside.

FACT

A conductor is a single wire, whereas a cable is composed of multiple wires under one protective means or layer.

Figure 4-5c. *Circular metal tubing is popular in commercial and light industrial areas.*

Figure 4-5d. *Circular nonmetallic tubing has become increasingly popular in commercial installations.*

an elementary school may start from an indoor electrical panel located in the school office. As the wiring method and conductors will be run above the office ceiling, EMT conduit (**Article 358**) is commonly used and permissible for this type of installation. As the conduit penetrates the outside wall and proceeds underground, it is to be routed under the bus stop canopy for a certain distance where it may be subject to physical damage. Because of the danger of physical damage, rules in **300.4** require conductors to be protected, so IMC (**Article 342)** is chosen. After clearing the bus stop space, the IMC conduit transitions to PVC (**Article 352**), commonly utilized for underground work, and continues underground to the sign location. The PVC travels underneath where the busses stop, so a burial depth of 24″ according to **Article 300.5** is used. At the sign, the conduit transitions again from PVC to IMC and continues out of the ground to the sign connection junction box.

Busways and cablebus are factory-assembled sections of grounded, completely enclosed, ventilated protective metal housings containing factory-mounted, bare or insulated conductors, which are usually copper or aluminum bars, rods, or tubes. **See Figure 4-6.**

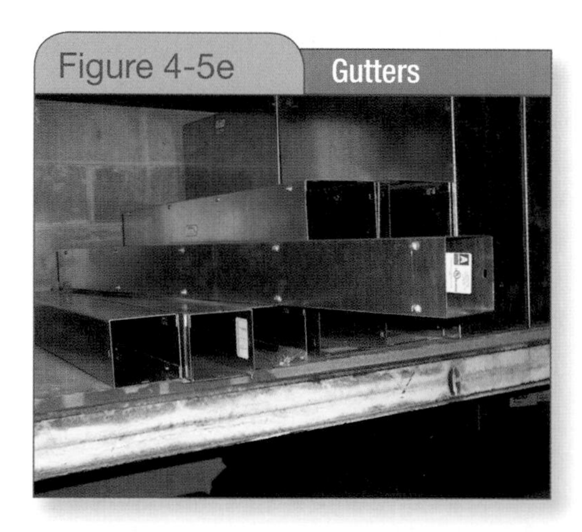

Figure 4-5e. *Non-circular raceways are especially used in industrial areas where a high quantity of conductors are employed.*

Figure 4-6. *Busways are used when a large current capacity is required and there is a need to supply multiple pieces of equipment within close proximity.*

Open wiring methods consist of concealed knob and tube wiring and open wiring on insulators. Note that these open wiring methods are without a protective outer jacket or enclosure and are extremely limited in application. This practice is not utilized in new installation but can be found in older maintenance situations.

WIRING MATERIALS

Wiring materials facilitate the installation of wiring methods. For example, enclosures and boxes are necessary for the installation of all types of wiring methods and equipment. Chapter 3 covers cabinets, cutout boxes, meter socket enclosures, outlet boxes, device boxes, pull boxes, junction boxes, conduit bodies, fittings, and handhole enclosures. **See Figure 4-7.**

Support systems are a type of wiring material. Messenger-supported wiring is a means of support for a given wiring method listed in **Article 396**. Cable trays are considered wiring materials and are not considered a wiring method. They are not listed as raceways in the *NEC*. Cable trays are utilized to support raceways, cables, and single conductors as covered in **Article 392**. **See Figure 4-8.**

GENERAL INFORMATION FOR ALL WIRING METHODS AND MATERIALS

Throughout the *NEC*, Articles and topics start generally and then move on to more specific topics. Chapter 3 does this by having the first article, **Article 300**, apply generally to wiring methods. As Chapter 3 continues, it includes more specific items such as enclosures, junction boxes, particular wiring methods, and raceways.

Article 300 Wiring Methods

The first article in Chapter 3, **Article 300 Wiring Methods**, covers the general rules for all wiring methods and wiring materials. All electrical installations must conform to these general requirements unless modified by another article in Chapter 5, 6, or 7. For example, **Articles 725** and **760** do not follow all of the requirements of **Article 300**, only those that are mentioned in **725** and **760**. See **725.3** and **760.3** for more detail. General requirements for all wiring methods and materials for each raceway or cable assembly article are listed in **Article 300** rather than being repeated in each individual section. Thus, **Article 300** is the backbone of the "Build" chapter. **Article 300** is divided into two logical parts as follows:

- General Requirements
- Requirements for over 1000 Volts

Review the following key provisions of **Article 300 Wiring Methods**:

300.3 Conductors requiring single conductors specified by *Code* are only allowed where part of a recognized wiring method of Chapter 3. These methods are generally categorized as cable assemblies, raceways, and open wiring. Conductors of different voltage systems 1000 volts or less, including AC and DC conductors, are allowed in the same raceway.

300.4 Protection Against Physical Damage requires that all conductors, raceways, and cables subject to physical damage be protected. Specific requirements are given for wiring methods through framing members and for ungrounded (hot) conductors of 4 AWG (American Wire Gage) or larger entering a cabinet, a box, an enclosure, or a raceway.

300.5 Underground Installations and **Table 300.5 Minimum Cover Requirements, 0 to 1000 Volts, Nominal, Burial in Millimeters (Inches)** contain specific requirements for all underground installations of direct-buried conductors, cables, and raceways.

300.11 Securing and Supporting requires that all raceways, cable assemblies, boxes, cabinets, and

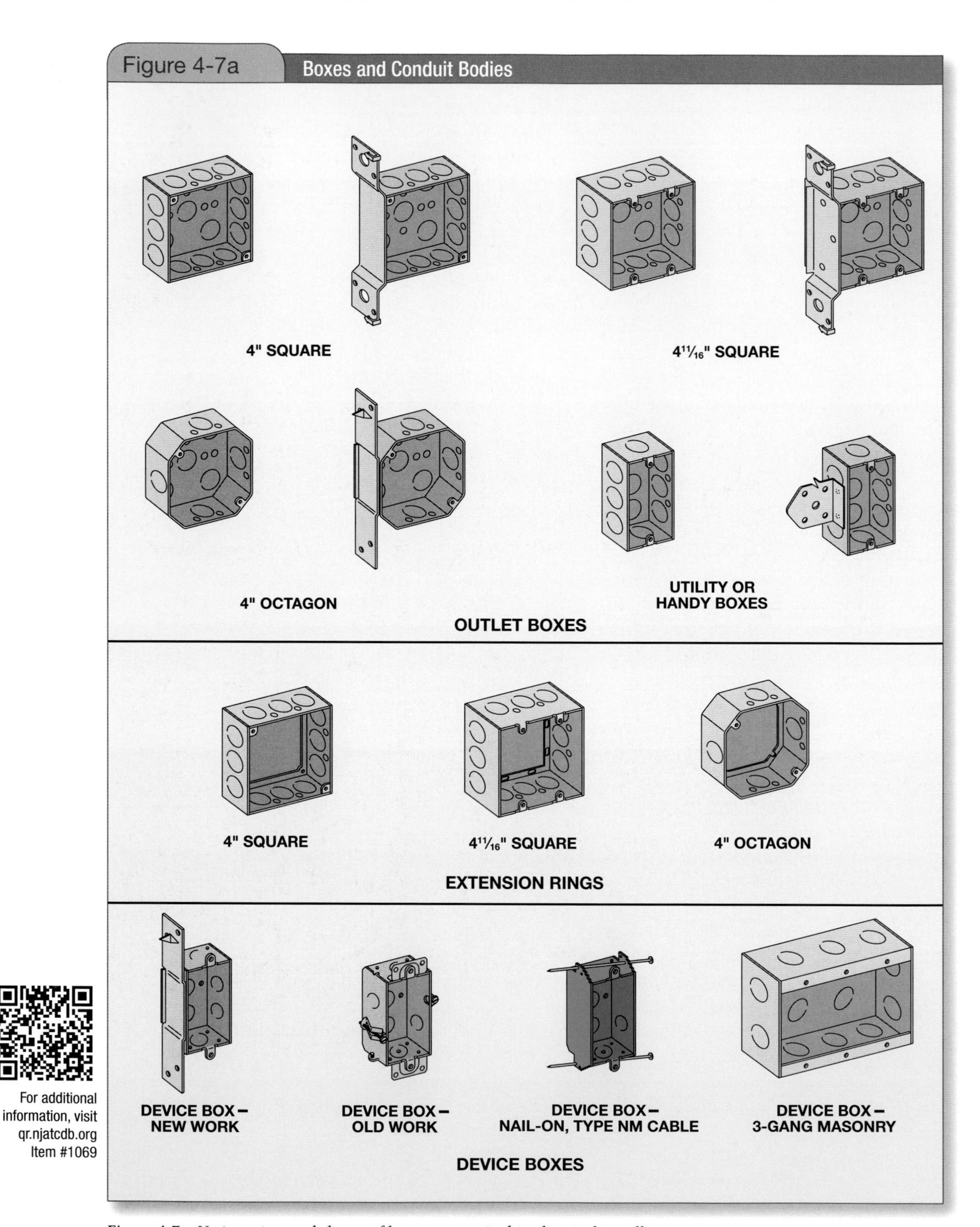

For additional information, visit qr.njatcdb.org Item #1069

Figure 4-7a. Various sizes and shapes of boxes are required in electrical installations.

Figure 4-7b Conduit Bodies

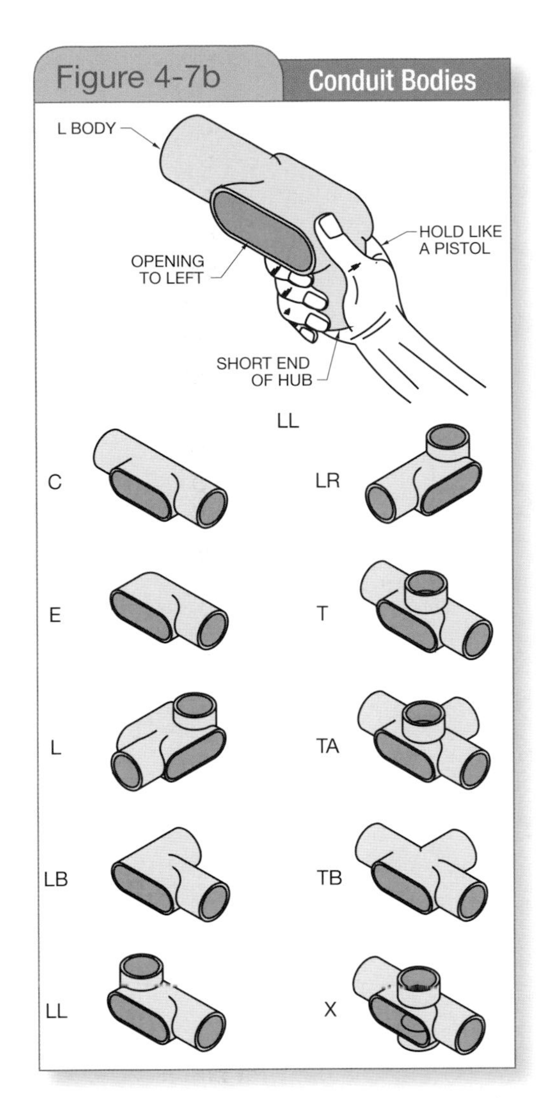

Figure 4-7b. Conduit bodies are required by the NEC to facilitate conductor installations.

Figure 4-8 Cable Tray

Figure 4-8. Cable trays come in various sizes and styles. This particular cable tray is often referred to as a ladder bottom tray. Many times the cable tray will have a solid bottom.

fittings be securely fastened in place. Additional requirements cover wiring located within floor, ceiling, and roof assemblies. Specific support requirements for raceways and for cable assemblies for support spacing are included in the respective raceway or cable assembly article.

300.14 Length of Free Conductors at Outlets, Junctions, and Switch Points requires a minimum length of conductor inside boxes or enclosures to allow for splicing or device termination. The conductor must be at least six inches long where it enters the box and also extend three inches beyond the opening of the box if the box or enclosure opening dimension is smaller than eight inches. The only exception to this requirement is if conductors are not spliced or terminated at the outlet, junction, or switch point.

For additional information, visit qr.njatcdb.org Item #1070

300.15 Boxes, Conduit Bodies, or Fittings—Where Required states that, in general, when raceways or cable assemblies are used, a box or conduit body (an LB, for example) must be installed at each conductor splice point and at each outlet, switch, junction, and termination point.

300.21 Spread of Fire or Products of Combustion requires that when wiring methods or materials penetrate an opening through fire-resistant walls, partitions, floors, or ceilings, the opening be installed with fire-stopped materials in accordance with approved methods to maintain their fire resistance rating.

300.22 Wiring in Ducts Not Used for Air Handling, Fabricated Ducts for Environmental Air, and Other Spaces for Environmental Air (Plenums) specifically addresses permitted wiring methods and materials in ducts, plenums, and other air-handling spaces.

Article 300 outlines general installation rules for all the wiring methods and materials listed in Chapter 3. *Code* users must become extremely familiar with the general requirements of **Article 300**, given their required application to every electrical installation regardless of the materials used.

Article 310 Conductors for General Wiring

All general requirements for conductors in cable assemblies, raceways, open wiring, or support systems are contained in **Article 310 Conductors for General Wiring.** Specific cables such as flexible cords are covered in **Article 400**, and specific fixture wires in **Article 402**, since flexible cords and fixture wires are not generally considered wiring methods. The use and installation rules are covered in Chapter 4; therefore they can be considered as "equipment for general use" and not wiring methods covered in Chapter 3.

Type designations, such as THHN, THWN or XHHW, are part of the required marking for all conductors. These letters and suffixes, used in the type designation to explain the physical properties of the conductor, are extremely important to the *Code* user. **See Figure 4-9.** The *NEC* discusses the labeling requirement to include: trade name, maximum operating temperatures, applications, insulation, thickness of insulation, and outer covering for the individual types of conductors. **See Figure 4-10.**

Insulation requirements for all conductors specify the type and thickness of all conductor insulation. Types of conductor insulation include, but are not limited to, TW, THW, THHW, THHN, RHH, RHW, and XHHW. The marking requirements for conductors ensure that all conductors and cables are marked to indicate the maximum rated voltage, the proper type designation, the manufacturer's name/trademark, and the size of the conductor

Figure 4-9 Conductor Type Designations

T	Thermoplastic insulation
R	Thermoset insulation
S	Silicone (thermoset) insulation
X	Cross-linked synthetic polymer insulation
Z	Modified tetrafluoroethylene insulation
U	Underground use
L	Lead sheath
N	Nylon jacket
W	Moisture resistant
H	75°C rated (lack of "H" usually indicates 60°C rating)
HH	90°C rated
-2	The suffix "-2" designates continuous 90°C rating, wet or dry

Figure 4-9. Letters and suffixes are used to explain the physical properties of a conductor.

Figure 4-10 Conductor Markings

*Figure 4-10. **Article 310** covers general requirements for conductors and their type designations, insulations, markings, mechanical strengths, ampacity ratings, and uses.*

in accordance with the American Wire Gage (AWG) or the circular mil area (cmil).

Mechanical strengths of conductors are addressed by requirements for minimum size, thickness of insulation, outer coverings, and permitted applications. Ampacity ratings for all types and uses of conductors are included in the ampacity tables in **Article 310**. Each of these tables addresses different installation possibilities for all types of conductors in all ranges of temperature limitations and installation methods. **Table 310.15(B)(16)**, however, is the most frequently used ampacity table. **310.15** provides specific requirements for conductor ampacity and the corrections for exceeding the number of current-carrying conductors permitted in the tables. Type designation letters, such as type THWN, are used to determine the ampacity of the conductor using the proper table.

Uses for all type designations of conductors are detailed in **Table 310.104(A)** and **Table 310.104(B)**. For example, type THHW is permitted for use in dry locations with a temperature limitation of 90°C, and in wet locations with a temperature limitation of 75°C. Notice the "W" in THHW meaning "wet location."

Code users must become familiar with the general requirements listed in **Article 310**, given their broad application to every electrical installation, regardless of the conductors used.

Cabinets, Boxes, Fittings, and Meter Socket/Handhole Enclosures

To facilitate the installation of all types of wiring methods, the need arises for wiring materials. Cabinets, boxes, fittings, and meter sockets, handholes, and enclosures are the wiring materials that are common to many electrical installations. The two articles in Chapter 3 that cover these wiring materials are **Article 312** and **Article 314**.

Article 312 is divided into two parts as follows:

- Installation
- Construction Specifications

Article 312 also contains important information regarding wiring space in panels, including space for splicing. **Article 312** is also referenced elsewhere in the code, in wireways (**Articles 376** and **378**) for requirements for proper sizing to allow enough room for the conductors to change direction in the wireway as needed.

Article 314 Outlet, Device, Pull, and Junction Boxes; Conduit Bodies; Fittings; and Handhole Enclosures

Article 314 provides general rules for the installation and construction of outlet, device, pull and junction boxes; conduit bodies; fittings; and handhole enclosures. **See Figure 4-11.**

Key provisions of **Article 314** include the following:

314.16 Number of Conductors in Outlet, Device, and Junction Boxes, and Conduit Bodies, along with the *Code* table provided, requires that all boxes and conduit bodies be of sufficient size to provide free space for enclosed conductors. Note that this section covers installations wherein all conductors enclosed are 6 AWG or smaller.

Figure 4-11 Handhole

Figure 4-11. Handhole boxes are utilized outdoors to provide pulling points and junction points for feeders and branch circuits. They are commonly found in landscaping, outdoor sports fields, and outdoor places where distances are great from the building and the outdoor equipment being served.

For additional information, visit qr.njatcdb.org Item #1071

Table 310.104(A) Conductor Applications and Insulations Rated 600 Volts[1]

Trade Name	Type Letter	Maximum Operating Temperature	Application Provisions	Insulation	Thickness of Insulation					Outer Covering[2]
					AWG or kcmil	mm		mils		
Fluorinated ethylene propylene	FEP or FEPB	90°C (194°F)	Dry and damp locations	Fluorinated ethylene propylene	14–10 8–2	0.51 0.76		20 30		None
		200°C (392°F)	Dry locations — special applications[3]	Fluorinated ethylene propylene	14–8	0.36		14		Glass braid
					6–2	0.36		14		Glass or other suitable braid material
Mineral insulation (metal sheathed)	MI	90°C (194°F) 250°C (482°F)	Dry and wet locations For special applications[3]	Magnesium oxide	18–16[4] 16–10 9–4 3–500	0.58 0.91 1.27 1.40		23 36 50 55		Copper or alloy steel
Moisture-, heat-, and oil-resistant thermoplastic	MTW	60°C (140°F) 90°C (194°F)	Machine tool wiring in wet locations Machine tool wiring in dry locations. Informational Note: See NFPA 79.	Flame-retardant, moisture-, heat-, and oil-resistant thermoplastic		(A)	(B)	(A)	(B)	(A) None (B) Nylon jacket or equivalent
					22–12 10 8 6 4–2 1–4/0 213–500 501–1000	0.76 0.76 1.14 1.52 1.52 2.03 2.41 2.79	0.38 0.51 0.76 0.76 1.02 1.27 1.52 1.78	30 30 45 60 60 80 95 110	15 20 30 30 40 50 60 70	
Paper		85°C (185°F)	For underground service conductors, or by special permission	Paper						Lead sheath
Perfluoro-alkoxy	PFA	90°C (194°F) 200°C (392°F)	Dry and damp locations Dry locations — special applications[3]	Perfluoro-alkoxy	14–10 8–2 1–4/0	0.51 0.76 1.14		20 30 45		None
Perfluoro-alkoxy	PFAH	250°C (482°F)	Dry locations only. Only for leads within apparatus or within raceways connected to apparatus (nickel or nickel-coated copper only)	Perfluoro-alkoxy	14–10 8–2 1–4/0	0.51 0.76 1.14		20 30 45		None
Thermoset	RHH	90°C (194°F)	Dry and damp locations		14-10 8–2 1–4/0 213–500 501–1000 1001–2000	1.14 1.52 2.03 2.41 2.79 3.18		45 60 80 95 110 125		Moisture-resistant, flame-retardant, nonmetallic covering[2]
Moisture-resistant thermoset	RHW RHW-2	75°C (167°F) 90°C (194°F)	Dry and wet locations	Flame-retardant, moisture-resistant thermoset	14–10 8–2 1–4/0 213–500 501–1000 1001–2000	1.14 1.52 2.03 2.41 2.79 3.18		45 60 80 95 110 125		Moisture-resistant, flame-retardant, nonmetallic covering
Silicone	SA	90°C (194°F) 200°C (392°F)	Dry and damp locations For special application[3]	Silicone rubber	14–10 8–2 1–4/0 213–500 501–1000 1001–2000	1.14 1.52 2.03 2.41 2.79 3.18		45 60 80 95 110 125		Glass or other suitable braid material

(continues)

Table 310.104(A) ***Continued***

Trade Name	Type Letter	Maximum Operating Temperature	Application Provisions	Insulation	Thickness of Insulation			Outer Covering[2]
					AWG or kcmil	mm	mils	
Thermoset	SIS	90°C (194°F)	Switchboard and switchgear wiring only	Flame-retardant thermoset	14–10 8–2 1–4/0	0.76 1.14 2.41	30 45 55	None
Thermoplastic and fibrous outer braid	TBS	90°C (194°F)	Switchboard and switchgear wiring only	Thermoplastic	14–10 8 6–2 1–4/0	0.76 1.14 1.52 2.03	30 45 60 80	Flame-retardant, nonmetallic covering
Extended polytetra-fluoro-ethylene	TFE	250°C (482°F)	Dry locations only. Only for leads within apparatus or within raceways connected to apparatus, or as open wiring (nickel or nickel-coated copper only)	Extruded polytetra-fluoroethylene	14–10 8–2 1–4/0	0.51 0.76 1.14	20 30 45	None
Heat-resistant thermoplastic	THHN	90°C (194°F)	Dry and damp locations	Flame-retardant, heat-resistant thermoplastic	14–12 10 8–6 4–2 1–4/0 250–500 501–1000	0.38 0.51 0.76 1.02 1.27 1.52 1.78	15 20 30 40 50 60 70	Nylon jacket or equivalent
Moisture- and heat-resistant thermoplastic	THHW	75°C (167°F) 90°C (194°F)	Wet location Dry location	Flame-retardant, moisture- and heat-resistant thermoplastic	14–10 8 6–2 1–4/0 213–500 501–1000 1001–2000	0.76 1.14 1.52 2.03 2.41 2.79 3.18	30 45 60 80 95 110 125	None
Moisture- and heat-resistant thermoplastic	THW	75°C (167°F) 90°C (194°F)	Dry and wet locations Special applications within electric discharge lighting equipment. Limited to 1000 open-circuit volts or less. (Size 14-8 only as permitted in 410.68.)	Flame-retardant, moisture- and heat-resistant thermoplastic	14–10 8 6–2 1–4/0 213–500 501–1000 1001–2000	0.76 1.14 1.52 2.03 2.41 2.79 3.18	30 45 60 80 95 110 125	None
	THW-2	90°C (194°F)	Dry and wet locations					
Moisture- and heat-resistant thermoplastic	THWN	75°C (167°F)	Dry and wet locations	Flame-retardant, moisture- and heat-resistant thermoplastic	14–12 10 8–6 4–2 1–4/0 250–500 501–1000	0.38 0.51 0.76 1.02 1.27 1.52 1.78	15 20 30 40 50 60 70	Nylon jacket or equivalent
	THWN-2	90°C (194°F)						
Moisture-resistant thermoplastic	TW	60°C (140°F)	Dry and wet locations	Flame-retardant, moisture-resistant thermoplastic	14–10 8 6–2 1–4/0 213–500 501–1000 1001–2000	0.76 1.14 1.52 2.03 2.41 2.79 3.18	3 45 60 80 95 110 125	None
Underground feeder and branch-circuit cable — single conductor (for Type UF cable employing more than one conductor, see Article 340).	UF	60°C (140°F) 75°C (167°F) [5]	See Article 340.	Moisture-resistant Moisture- and heat-resistant	14–10 8–2 1–4/0	1.52 2.03 2.41	60[6] 80[6] 95[6]	Integral with insulation

(continues)

Table 310.104(A) ***Continued***

Trade Name	Type Letter	Maximum Operating Temperature	Application Provisions	Insulation	Thickness of Insulation: AWG or kcmil	mm	mils	Outer Covering[2]
Underground service-entrance cable — single conductor (for Type USE cable employing more than one conductor, see Article 338).	USE	75°C (167°F)[5]	See Article 338.	Heat- and moisture-resistant	14–10 8–2 1–4/0 213–500 501–1000 1001–2000	1.14 1.52 2.03 2.41 2.79 3.18	45 60 80 95[7] 110 125	Moisture-resistant nonmetallic covering (See 338.2.)
	USE-2	90°C (194°F)	Dry and wet locations					
Thermoset	XHH	90°C (194°F)	Dry and damp locations	Flame-retardant thermoset	14–10 8–2 1–4/0 213–500 501–1000 1001–2000	0.76 1.14 1.40 1.65 2.03 2.41	30 45 55 65 80 95	None
Thermoset	XHHN	90°C (194°F)	Dry and damp locations	Flame-retardant thermoset	14–12 10 8–6 4–2 1–4/0 250–500 501–1000	0.38 0.51 0.76 1.02 1.27 1.52 1.78	15 20 30 40 50 60 70	Nylon jacket or equivalent
Moisture-resistant thermoset	XHHW	90°C (194°F)	Dry and damp locations	Flame-retardant, moisture-resistant thermoset	14–10 8–2 1–4/0 213–500 501–1000 1001–2000	0.76 1.14 1.40 1.65 2.03 2.41	30 45 55 65 80 95	None
		75°C (167°F)	Wet locations					
Moisture-resistant thermoset	XHHW-2	90°C (194°F)	Dry and wet locations	Flame-retardant, moisture-resistant thermoset	14–10 8–2 1–4/0 213–500 501–1000 1001–2000	0.76 1.14 1.40 1.65 2.03 2.41	30 45 55 65 80 95	None
Moisture-resistant thermoset	XHWN	75°C (167°F)	Dry and wet locations	Flame-retardant, moisture-resistant thermoset	14–12 10 8–6 4–2 1–4/0 250–500 501–1000	0.38 0.51 0.76 1.02 1.27 1.52 1.78	15 20 30 40 50 60 70	Nylon jacket or equivalent
	XHWN-2	90°C (194°F)						
Modified ethylene tetrafluoro-ethylene	Z	90°C (194°F)	Dry and damp locations	Modified ethylene tetrafluoro-ethylene	14–12 10	0.38 0.51	15 20	None
		150°C (302°F)	Dry locations — special applications[3]		8–4 3–1 1/0–4/0	0.64 0.89 1.14	25 35 45	
Modified ethylene tetrafluoro-ethylene	ZW	75°C (167°F)	Wet locations	Modified ethylene tetrafluoro-ethylene	14–10 8–2	0.76 1.14	30 45	None
		90°C (194°F)	Dry and damp locations					
		150°C (302°F)	Dry locations — special applications[3]					
	ZW-2	90°C (194°F)	Dry and wet locations					

[1]Conductors can be rated up to 1000 V if listed and marked.
[2] Some insulations do not require an outer covering.
[3] Where design conditions require maximum conductor operating temperatures above 90°C (194°F).
[4] For signaling circuits permitting 300-volt insulation.
[5] For ampacity limitation, see 340.80.
[6] Includes integral jacket.
[7] Insulation thickness shall be permitted to be 2.03 mm (80 mils) for listed Type USE conductors that have been subjected to special investigations. The nonmetallic covering over individual rubber-covered conductors of aluminum-sheathed cable and of lead-sheathed or multiconductor cable shall not be required to be flame retardant. For Type MC cable, see 330.104. For nonmetallic-sheathed cable, see Article 334, Part III. For Type UF cable, see Article 340, Part III.

Reprinted with permission from NFPA 70-2017, *National Electrical Code*®, Copyright© 2016, National Fire Protection Association, Quincy, MA 02169. This reprinted material is not the complete and official position of the NFPA on the referenced subject, which is represented only by the standard in its entirety.

314.23 Supports provide minimum requirements for the support of all boxes and enclosures. A careful review of this section is important because its requirements are utilized in nearly every electrical installation.

314.28 Pull and Junction Boxes and Conduit Bodies provides minimum requirements for the size of all pull and junction boxes and conduit bodies. Note that this section covers installations wherein conductors 4 AWG or larger are enclosed.

314.29 Boxes, Conduit Bodies, and Handhole Enclosures to Be Accessible prohibits boxes and conduit bodies from being concealed. Note that a box and raceway installed above a lay-in type drop ceiling would be considered exposed work and accessible.

Article 314 is divided into four logical parts as follows:

Part I. Scope and General
Part II. Installation
Part III. Construction Specifications
Part IV. Pull and Junction Boxes, Conduit Bodies, and Handhole Enclosures for Use on Systems over 1000 Volts, Nominal

Table 310.104(B) Thickness of Insulation for Nonshielded Types RHH and RHW Solid Dielectric Insulated Conductors Rated 2000 Volts

Conductor Size (AWG or kcmil)	Column A[1]		Column B[2]	
	mm	mils	mm	mils
14–10	2.03	80	1.52	60
8	2.03	80	1.78	70
6–2	2.41	95	1.78	70
1–2/0	2.79	110	2.29	90
3/0–4/0	2.79	110	2.29	90
213–500	3.18	125	2.67	105
501–1000	3.56	140	3.05	120
1001–2000	3.56	140	3.56	140

[1]Column A insulations are limited to natural, SBR, and butyl rubbers.
[2]Column B insulations are materials such as cross-linked polyethylene, ethylene propylene rubber, and composites thereof.

THE *NEC* PARALLEL NUMBERING SYSTEM – CHAPTER 3

All cable assembly and circular raceway articles share a common article layout and section numbering system, or what some call the "Parallel Numbering System," intended to provide the *Code* user with a more consistent, easy-to-use format. Many of the "other than circular" raceways have also adopted this common format. Once familiar with the common format of these articles, the *Code* user can quickly and accurately move through the many different wiring methods in Chapter 3 to find the needed information and requirements. For example, **Section XXX.10** of each article is "Uses Permitted." A *Code* user familiar with this common numbering scheme could quickly and accurately determine the wiring methods permitted for a particular installation.

Within the common format is included specific section numbering, which may not apply to all wiring methods. For example, **Section 3XX.28** is reserved for "Reaming and Threading" or "Trimming" and can be found only in raceway articles where these installation procedures are required.

The common article layout and section numbering is as follows:

Article 3XX
Part I. General
3XX.1 Scope
3XX.2 Definition(s)
3XX.3 Other Articles
3XX.6 Listing Requirements

Part II. Installation
3XX.10 Uses Permitted
3XX.12 Uses Not Permitted
3XX.15 Exposed Work
3XX.17 Through or Parallel to Framing Members
3XX.19 Clearances
3XX.20 Size
3XX.22 Number of Conductors
3XX.23 Inaccessible Attics
3XX.24 Bends – How Made
3XX.26 Bends – Number in One Run

3XX.28 Reaming and Threading (For metallic raceways)
3XX.28 Trimming (For non-metallic raceways)
3XX.30 Securing and Supporting
3XX.40 Boxes and Fittings
3XX.42 Couplings, Connectors, Devices
3XX.44 Expansion Fittings
3XX.46 Bushings
3XX.48 Joints
3XX.56 Splices and Taps
3XX.60 Grounding, Bonding
3XX.80 Ampacity

Part III. Construction Specifications
3XX.100 Construction
3XX.104 Conductors
3XX.108 Equipment Grounding Conductor
3XX.112 Insulation
3XX.116 Sheath, Jacket, Conduit
3XX.120 Marking(s)
3XX.130 Standard Lengths

Cable Assemblies

Chapter 3 recognizes 11 types of cable assemblies as acceptable wiring methods. All cable assembly articles are listed in alphabetical order as follows:

Article 320 Armored Cable: Type AC
Article 322 Flat Cable Assemblies: Type FC

Figure 4-12 FMC (Article 348)

Figure 4-12. Flexible metal conduit can be used to connect electrical equipment that vibrates, or to connect equipment such as light fixtures in lay-in ceilings.

Article 324 Flat Conductor Cable: Type FCC
Article 326 Integrated Gas Spacer Cable: Type IGS
Article 328 Medium Voltage Cable: Type MV
Article 330 Metal-Clad Cable: Type MC
Article 332 Mineral-Insulated, Metal-Sheathed Cable: Type MI
Article 334 Nonmetallic-Sheathed Cable: Types NM, NMC, and NMS
Article 336 Power and Control Tray Cable: Type TC
Article 338 Service-Entrance Cable: Types SE and USE
Article 340 Underground Feeder and Branch-Circuit Cable: Type UF

Raceways, Circular Metal Conduit

Four types of circular metal conduits are recognized as an acceptable wiring method in Chapter 3. They are defined as rigid-type conduits or flexible-type conduits as follows:

Article 342 Intermediate Metal Conduit: Type IMC
Article 344 Rigid Metal Conduit: Type RMC

NOTE: RMC includes galvanized steel, stainless steel, brass, aluminum, and ferrous.

Article 348 Flexible Metal Conduit: Type FMC. See Figure 4-12.
Article 350 Liquidtight Flexible Metal Conduit: Type LFMC

Raceways, Circular Nonmetallic Conduit

Five types of circular nonmetallic conduits are recognized as an acceptable wiring method in Chapter 3. They are defined as rigid-type conduits or flexible-type conduits as follows:

Article 352 Rigid Polyvinyl Chloride Conduit: Type PVC
Article 353 High-Density Polyethylene Conduit: Type HDPE Conduit

Article 354 Nonmetallic Underground Conduit with Conductors: Type NUCC
Article 355 Reinforced Thermosetting Resin Conduit: Type RTRC
Article 356 Liquidtight Flexible Nonmetallic Conduit: Type LFNC

Raceways, Circular Metallic Tubing

Chapter 3 recognizes two types of circular metallic tubing as acceptable wiring methods. Note that electrical metallic tubing, type EMT, is commonly referred to as "thinwall conduit" but is designated as "tubing" in the *NEC*. The two types of tubing, a single rigid type, and a single flexible type, are covered in the following articles:

Article 358 Electrical Metallic Tubing: Type EMT
Article 360 Flexible Metallic Tubing: Type FMT

Raceways, Circular Nonmetallic Tubing

Chapter 3 recognizes a single type of circular nonmetallic tubing as an acceptable wiring method.

Article 362 Electrical Nonmetallic Tubing: Type ENT

Factory-Assembled Power Distribution Systems

The *NEC* recognizes two types of factory-assembled power distribution systems as acceptable wiring methods in Chapter 3. These wiring methods, busway, and cablebus, are preassembled and bolted together for a complete installation in the field. These systems allow for a disconnecting means and overcurrent protection to be installed anywhere along the installation of busway and cablebus, providing an easy means for power distribution. The two articles for busway and cablebus numerically separate the "Circular Raceways" from the "Other than Circular Raceways," as follows:

Article 368 Busways
Article 370 Cablebus

Raceways Other Than Circular

Ten types of other-than-circular raceways are recognized as an acceptable wiring method in Chapter 3:

Article 366 Auxiliary Gutters
Article 372 Cellular Concrete Floor Raceways
Article 374 Cellular Metal Floor Raceways
Article 376 Metal Wireways
Article 378 Nonmetallic Wireways
Article 380 Multioutlet Assembly
Article 384 Strut-Type Channel Raceway
Article 386 Surface Metal Raceways
Article 388 Surface Nonmetallic Raceways
Article 390 Underfloor Raceways

Surface-Mounted Nonmetallic Branch-Circuit Extension

Article 382 shows two methods to extend a branch circuit. The first method is primarily limited to use only from an existing outlet in a residential or commercial occupancy not more than three floors above grade. It was used primarily in older electrical installations to allow for the surface mounting of additional receptacle outlets. The second method has provisions to allow a "concealable nonmetallic extension" to be installed on walls or ceilings covered with paneling, tile, paint, joint compound, or similar material.

Support Systems for Cables/Raceways

Two types of systems are recognized as acceptable for the support of wiring methods covered in Chapter 3:

Article 392 Cable Trays permits, under specified conditions, cable trays to support single conductors, cable assemblies, and raceways.
Article 396 Messenger-Supported Wiring permits, as the name implies, an exposed wiring support system using a messenger wire to

support insulated conductors. This system is permitted to support only those cable assemblies or conductors listed in **Table 396.10(A)**.

Open-Type Wiring Methods

Three types of open-type wiring methods are permitted in Chapter 3, although they are of extremely limited use:

Article 394 Concealed Knob-and-Tube Wiring
Article 398 Open Wiring on Insulators
Article 399 Outdoor Overhead Conductors over 1000 Volts

KEY WORDS AND CLUES FOR CHAPTER 3, "BUILD"

Wiring Methods

- General questions for wiring method installation
- Conductors, types, uses, ampacity
- Cable assemblies, all types
- Conduits, all types
- Tubing, all types
- Other raceways, all types
- Installation of all wiring methods
- Support of all wiring methods
- Construction of all wiring methods
- Uses permitted or not permitted, all wiring methods

Wiring Materials

- General questions for wiring materials
- Cabinets
- Cutout boxes
- Meter socket enclosures
- Outlet, device, pull and junction boxes
- Conduit bodies
- Handhole enclosures
- Support systems, cable tray, and messenger-supported wiring
- Construction of wiring materials
- Installation of wiring materials
- Support of wiring materials

If an inquiry includes any wiring and/or materials used with wiring, think "BUILD" and go to Chapter 3. Refer to Chapter 3 when a question or need within the *NEC* deals with any of the following:

- Wiring methods, cable assemblies, raceways
- Conductors, type designation, use, ampacity
- Installation of wiring methods
- Wiring materials, enclosures of all types
- Installation, use, and size of wiring materials
- Any question on the physical installation of wiring methods and materials
- Support systems for raceways, conductors, or cable assemblies

Summary

Chapter 3, in accordance with **Section 90.3**, applies generally to all electrical installations. Using the *Codeology* method, this chapter has been called the "Build" chapter due to the scope of its coverage. Chapter 3 is a "hands-on" and mechanical chapter, in that all of the material covered is to be physically installed. It is the means by which electrical current is delivered from the source of power to the last outlet in the electrical distribution system.

Summary

From the *NEC* title for **Chapter 3 Wiring Methods and Materials**, the scope of this chapter can be described as "Information and Rules on Wiring Methods and Materials for Electrical Installations." Chapter 3 covers the entire electrical distribution system, from the service point (connection to the utility) to the last receptacle or other outlet in the electrical system. All of the wiring methods and materials used to distribute electrical energy from the source to the last outlet are covered in Chapter 3. In accordance with its scope of "Wiring Methods," Chapter 3 provides detailed requirements for all conductors, raceways, cable assemblies, and other recognized wiring methods. In accordance with the scope of "Wiring Materials," it provides detailed requirements for all enclosures, boxes, conduit bodies, and support systems.

Chapter 3, along with Chapters 1, 2, and 4, builds the foundation or backbone of all electrical installations.

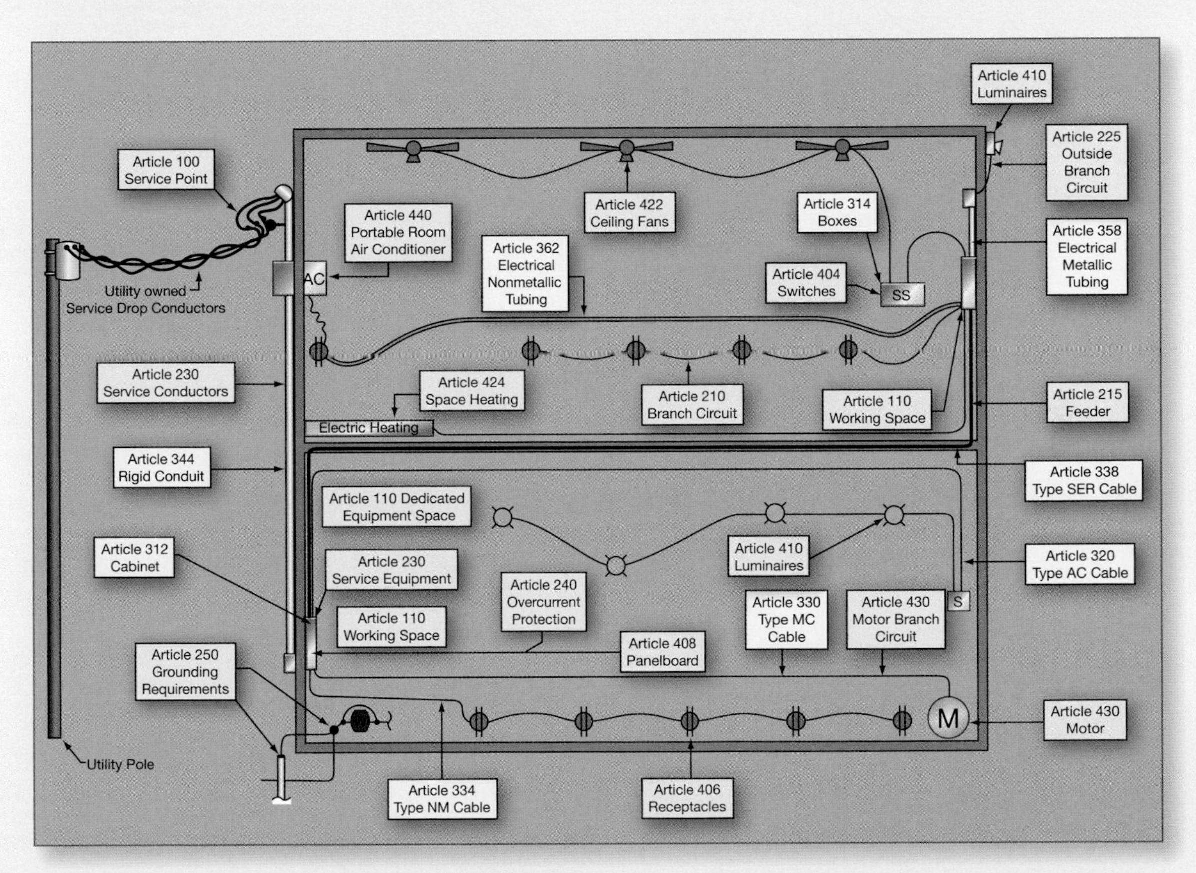

Chapter 3 applies generally in all electrical installations.

Review Questions

1. **Chapter 3 of the *NEC* applies to all electrical installations covered by the *NEC*.**
   a. True
   b. False
2. **Chapter 3 is subdivided into how many articles?**
   a. 39
   b. 41
   c. 44
   d. 46
3. **The scope of *NEC* Chapter 3 is divided into two areas covering wiring devices and wiring materials.**
   a. True
   b. False
4. **Name the part and article of Chapter 3 that addresses the installation of a pull box containing conductors rated at 13,200 volts.**
   a. **Article 300 Part II**
   b. **Article 356 Part II**
   c. **Article 394 Part III**
   d. **Article 399**
5. **Chapter 3 of the *NEC* addresses wiring methods, which would include circular raceways. Which article would apply to the installation of type RMC conduit?**
   a. **Article 328**
   b. **Article 342**
   c. **Article 344**
   d. **Article 352**
6. **300.22(C) provides information to aid the *Code* user in understanding what "other space" is in relation to a duct or plenum.**
   a. True
   b. False
7. **Which two articles in *NEC* Chapter 3 cover requirements for support systems for cables and/or raceways?**
   a. **Articles 320** and **330**
   b. **Articles 342** and **344**
   c. **Articles 360** and **366**
   d. **Articles 392** and **396**

Review Questions

8. Article 340 is divided into how many parts?

a. 2
b. 3
c. 4
d. 5

9. Part III of Article 340 applies to the installation of type UF cable.

a. True
b. False

10. Which two articles in Chapter 3 cover requirements for factory-assembled power distribution systems?

a. **Articles 300** and **310**
b. **Articles 326** and **328**
c. **Articles 368** and **370**
d. **Articles 392** and **393**

11. Chapter 3 is known in generic *Codeology* language as __?__.

a. Build
b. Plan
c. Specials
d. Use

The *NEC* "Use" Chapter (Equipment for General Use)

Introduction

Chapter 4 of the *NEC*, the "Use" chapter, provides rules and information on electrical equipment for general use. Any "Special Equipment" that requires unique installations rules is addressed in Chapter 6, in accordance with **Section 90.3**.

Chapter 2 of the *NEC* "Plans" for the installation of general electrical connections of systems and devices and provides information regarding wiring systems and protection. Chapter 3 "Builds" a general electrical installation by delivering electrical energy from the source to the load(s). All equipment covered in Chapter 4 is dedicated to the "Use" of electrical energy and addresses the control of electrical energy through devices or consumption of electrical energy by utilization equipment.

Electrical devices are used to provide power to cord- and plug-connected equipment through receptacles, switch lighting and other loads, and control other types of electrical equipment that uses electrical energy, performs a task, or provides a service for the consumer. For example, motors are used in several different applications; electrical space heaters provide heat; air conditioners cool homes; voltage is transformed from one level to another; and sometimes other sources of power are used in addition to utility power, such as wind generators or photovoltaic systems, generators, and batteries.

Objectives

» Associate the *Codeology* title for *NEC* Chapter 4 as "Use."

» Identify the type of information and requirements dealing with the installation, control, and supply for utilization equipment covered in *NEC* Chapter 4.

» Recognize Chapter 4 numbering as the 400-series.

» Recognize, recall, and become familiar with articles contained in Chapter 4.

Chapter 5

Table of Contents

NEC CHAPTER 4, "USE"

Although Chapter 4 of the *NEC* is identified as the "Use" chapter, not all of the equipment it covers uses electrical energy. The term "Use" can have an alternate meaning, as many of the articles in Chapter 4 cover devices and equipment frequently "used" after the building is turned over to the owner. However the term is used, all of the articles in Chapter 4 play a major role in the use of electrical energy for accomplishing a desired task.

These examples illustrate the makeup of Chapter 4, the "Use" chapter:

- Flexible cords and cables allow for the connection of appliances and other utilization equipment to an electrical outlet.
- Fixture wires provide for the wiring of luminaires.
- Receptacle outlets facilitate the use of appliances and other loads.
- Switches control lighting and all other loads.
- Luminaires (lighting fixtures), appliances, heating equipment, motors, and air-conditioning/refrigeration equipment all use electrical energy.
- Panelboards, switchboards, industrial control panels, and switches provide control and overcurrent protection for all conductors supplying end-use equipment.
- Generators provide a power source for emergency, legally required standby and optional standby systems. Other sources of power, such as solar, wind, and fuel cell systems are considered special and are covered by Chapter 6 of the *NEC*.
- Transformers enable the use of electrical energy and provide the flexibility to create a new system to allow the consumption of electrical energy at utilization voltages. For example, a service may be 277/480 volts, 3-phase, 4-wire to supply air-conditioning and refrigeration equipment. A transformer is installed that creates a new system at 120/208 volts to allow electrical energy consumption at 120 volts for general receptacle outlets.
- Phase converters, capacitors, resistors, and reactors allow for the economical use of electrical energy.

There are 22 articles in Chapter 4, the 400-series. These address equipment for general use to facilitate the utilization of electrical energy. The requirements and information in this chapter are logically divided into seven categories. **See Figure 5-1.**

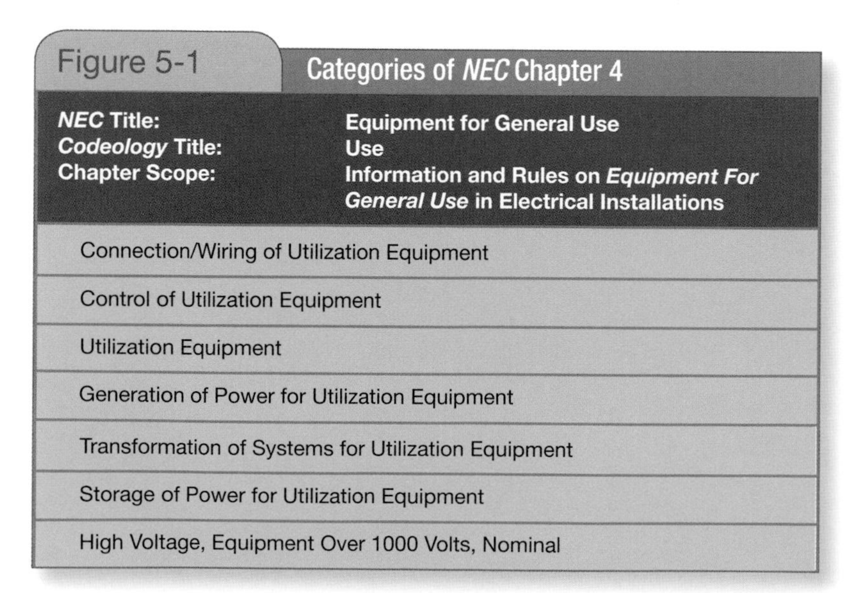

Figure 5-1 — Categories of *NEC* Chapter 4

***NEC* Title:**	**Equipment for General Use**
***Codeology* Title:**	**Use**
Chapter Scope:	**Information and Rules on *Equipment For General Use* in Electrical Installations**
Connection/Wiring of Utilization Equipment	
Control of Utilization Equipment	
Utilization Equipment	
Generation of Power for Utilization Equipment	
Transformation of Systems for Utilization Equipment	
Storage of Power for Utilization Equipment	
High Voltage, Equipment Over 1000 Volts, Nominal	

Figure 5-1. Chapter 4 is broken into seven categories facilitating the utilization of electrical energy.

CATEGORIZATION OF CHAPTER 4

When using the *Codeology* method, the title for Chapter 4 is "Use." This title encompasses all equipment that uses electrical energy as well as all associated equipment necessary to safely accomplish utilization. To further explain, equipment that uses electricity includes luminaires (light fixtures), different kinds of appliances, electric heating equipment, motors, and air conditioning. Equipment that safely accomplishes this utilization of power, or distributes electrical energy, includes cords, fixture wires, receptacles, switches, switchboards and panelboards, transformers, and phase converters. Auxiliary equipment associated with motors and transformers includes capacitors, resistors, and reactors. Additional equipment in Chapter 4 that produces and also provides for the utilization of electricity includes generators and batteries. **See Figure 5-2.**

CONNECTION AND WIRING OF UTILIZATION EQUIPMENT

Flexible cords and cables are appropriately located in Chapter 4, the "Use" chapter. Flexible cords and cables are not permitted as a substitute for fixed wiring. This means that while cords and cables are permitted

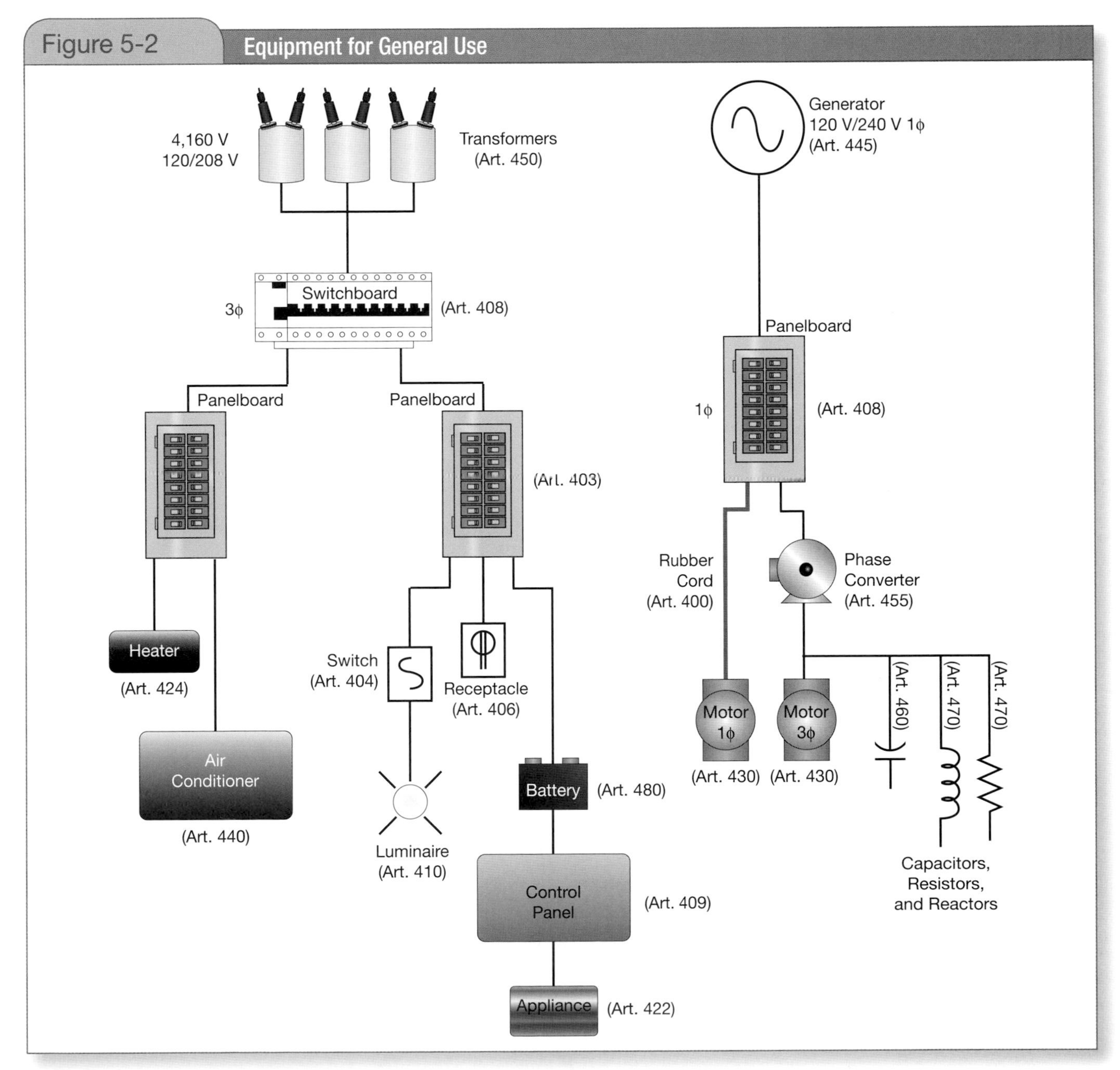

Figure 5-2. Chapter 4 of the NEC *covers Equipment for General Use such as transformers, panels, motors, heaters, switches, receptacles, and flexible cords.*

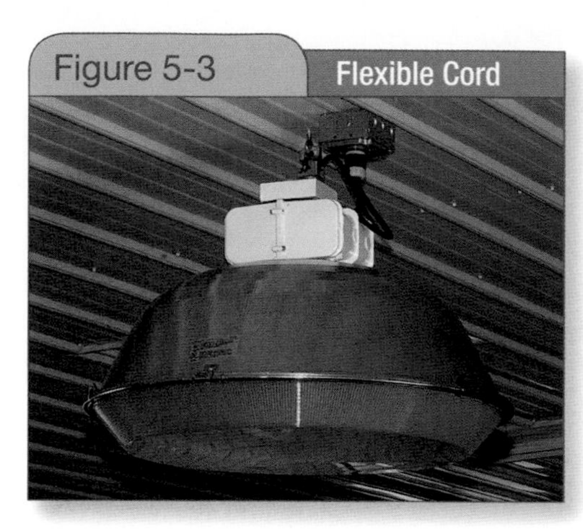

Figure 5-3. Typically, luminaires are supplied with power by flexible cords and cables.

to supply utilization equipment, they are not permitted as "wiring methods." A good example of a flexible cord is an extension cord used daily at construction sites. Chapter 3 of the *NEC* is dedicated to wiring methods and wiring materials. Flexible cord and cables facilitate the installation of equipment that is frequently changed or partially mobile. **See Figure 5-3.**

The *NEC* uses the widely known international term for lighting fixtures, *luminaires*. **Article 402** provides information and requirements for the use and limits of fixture wire in luminaires and associated equipment. It is important to note that fixture wires are not permitted to serve as branch circuits. They are permitted for installation in lighting fixtures or associated equipment. As covered in **Article 310**, fixture wires are rated at higher temperatures than normal general use conductors due to the heat produced in most luminaire fixtures. Fixture wires are also allowed for Class 1 circuits in **Article 725**, and fire alarm circuits in **Article 760.**

Electrical systems are installed in all occupancies to allow for the use of electrical equipment. Utilization equipment must be connected to a branch circuit through hardwiring or cord-and-plug connection. **Article 400** and **Article 402** provide information and requirements on the use of cords and cables and fixture wires.

ARTICLE 404 SWITCHES – CONTROL OF UTILIZATION EQUIPMENT

Switches are appropriately located in Chapter 4, the "Use" chapter. Switches are essential to the control and protection (fused switches) of conductors and utilization equipment. Receptacles, cord connectors, and attachment plugs are also covered by Chapter 4, the "Use" chapter. The connection and disconnection of appliances and other cord-and-plug-connected utilization equipment would not be possible without these devices. In addition, switchboards, panelboards, and industrial control panels are also covered by Chapter 4 and play an essential part to allow for control and protection of all conductors and utilization equipment.

The *NEC* requires control and protection of all conductors and equipment in an electrical installation. In some cases, utilization equipment is controlled and protected through the same devices used for overcurrent protection of branch-circuit conductors. However, not all electrical "equipment for general use" is located at the end of the branch circuit with the utilization equipment.

Article 404 Switches covers switches and devices such as circuit breakers when used as switches, while **Article 406 Receptacles, Cord Connectors, and Attachment Plugs (Caps)** covers the devices providing control and flexibility. **See Figure 5-4.**

Article 408 Switchboards and Panelboards and **Article 409 Industrial Control Panels** provide provisions for the control and protection of utilization equipment. In most cases, this occurs at the source of the feeder and branch circuits supplying the equipment. Switchboards typically provide distribution to larger loads and other panels. **See Figure 5-5.** Panelboards are typically full of 15- and 20-amp breakers feeding branch circuits of receptacles and luminaires. **See Figure 5-6.**

Figure 5-4 Receptacles and Switches for General Use

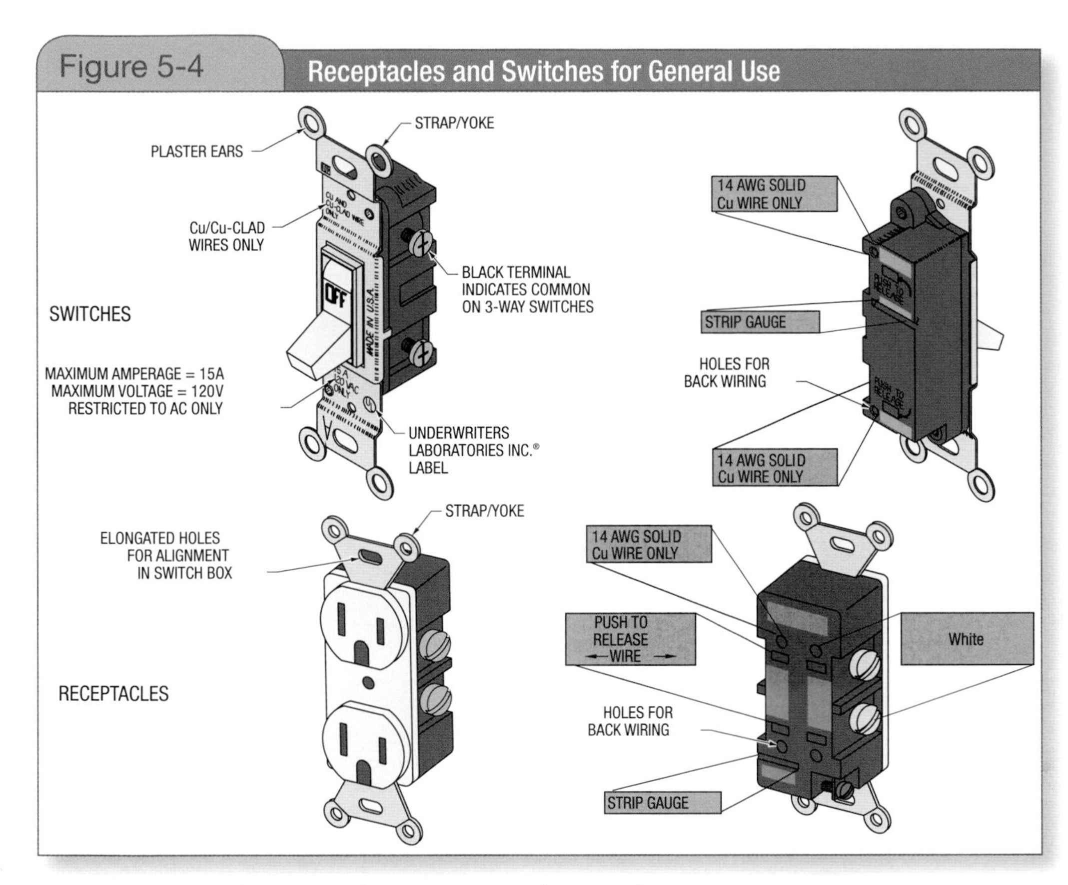

Figure 5-4. Receptacles and switches are equipment for general use.

Figure 5-5 Switchboards

Figure 5-5. Switchboards typically supply service to larger loads such as motors, transformers, and larger-ampacity equipment.

Figure 5-6 Panelboards

Figure 5-6. Panelboards are typically used for branch circuits to supply wiring devices such as receptacles.

Figure 5-7a Luminaires

Figure 5-7a. ***Article 410*** *covers the installation of luminaires, while* ***Article 422*** *spells out the requirements for typical appliances such as a ceiling fan.*

Figure 5-7b Fixed Heaters

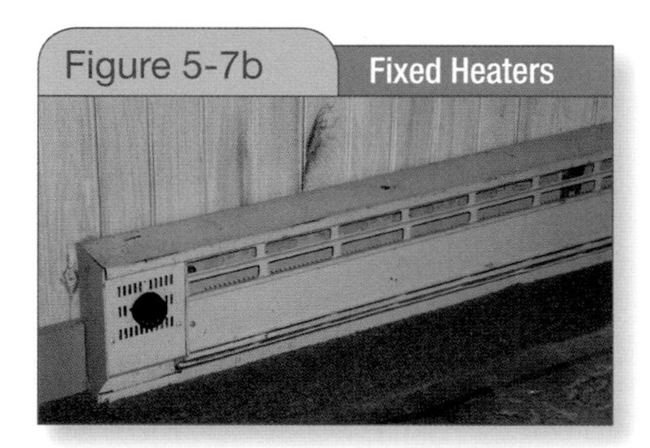

Figure 5-7b. ***Article 424*** *provides details of installation for fixed space heating equipment.*

Figure 5-7c Motors

Figure 5-7c. ***Article 430*** *lists the installation requirements of motors.*

Figure 5-7d Air-Conditioning

Figure 5-7d. ***Article 440*** *covers the installation of air-conditioning and refrigeration equipment.*

Categories of Utilization Equipment

Luminaires (lighting fixtures), appliances, space-heating equipment, deicing and snow-melting equipment, heat trace, motors, AC, and refrigeration equipment are all appropriately located in Chapter 4, the "Use" chapter. This equipment uses electrical energy. **See Figure 5-7.**

Eight articles in the *NEC* specifically address the requirements for utilization equipment. These requirements include those for general coverage, installation, control and protection, disconnecting means, construction, and many other equipment-specific needs. The *NEC* requirements for special utilization equipment are located in Chapter 6, in conformance with **Section 90.3**. General utilization equipment in Chapter 4 of the *NEC* is divided into four basic categories: lighting, appliances, heating, and motors.

- **Article 410 Luminaires, Lampholders, and Lamps**
- **Article 411 Lighting Systems Operating at 30 Volts or Less**
- **Article 422 Appliances**
- **Article 424 Fixed Electric Space-Heating Equipment**

- **Article 425 Fixed Resistance and Electrode Industrial Process Heating Equipment**
- **Article 426 Fixed Outdoor Electric Deicing and Snow-Melting Equipment**
- **Article 427 Fixed Electric Heating Equipment for Pipelines and Vessels**
- **Article 430 Motors, Motor Circuits, and Controllers**
- **Article 440 Air-Conditioning and Refrigeration Equipment**

Generation of Power for Utilization Equipment

Coverage of generators is appropriately located in Chapter 4, the "Use" chapter. In almost all electrical installations, the power is supplied to an occupancy by an electric utility and is called a service. When the need for an emergency or standby system arises, the most common or general backup system is an on-site standby generator. **See Figure 5-8.**

Article 445 provides information and requirements for generators, which are considered equipment for general use. In accordance with **Section 90.3**, the *NEC* addresses two additional "special" energy systems in Chapter 6. **Article 445 Generators** addresses the generation requirements of Chapter 4 and is not divided into parts.

Figure 5-8 Generator

Figure 5-8. Generators are necessary in many types of installations to supply power for utilization equipment. Generators mmay be mounted on top of a day fuel tank.

Transformation of Systems for Utilization Equipment

Transformers are appropriately located in Chapter 4, the "Use" chapter. When utilization equipment operates at different system voltages in an electrical installation, transformers are used to derive new systems at the utilization voltage to meet the system requirements.

Transformers are appropriately located in Chapter 4, the "Use" chapter. When utilization equipment operates at different system voltages in an electrical installation, transformers are used to derive new systems at the utilization voltage to meet the system requirements.

Phase converters, capacitors, resistors, and reactors are also covered in Chapter 4 and allow economical and efficient use of utilization equipment. For example, consider an older building or structure with a single-phase, 3-wire service. To use a 3-phase motor in this occupancy, a phase converter would be required. In addition, power factor problems of wasted energy and high cost can be corrected through the application of capacitors.

Article 450 covers transformers, which are an essential part of most electrical installations. They provide an installation with the flexibility of deriving a new system to meet the requirements of utilization equipment. Phase converters, covered by **Article 455**, allow single-phase systems to derive a 3-phase system and older 2-phase electrical systems to utilize the 3-phase equipment, which is more common, less expensive, and readily available. While capacitors, resistors, and reactors do not transform or change a system, they are essential equipment

Figure 5-9a. Pad-mounted transformers typically feed the entire building with power.

Figure 5-9b. Dry-type transformers are usually mounted inside a building and provide power for specific loads or an additional voltage service.

that supplements electrical systems in special applications.

Four articles address the transformation or adjustment of system requirements of Chapter 4:

- **Article 450 Transformers and Transformer Vaults (Including Secondary Ties)**
- **Article 455 Phase Converters**
- **Article 460 Capacitors**
- **Article 470 Resistors and Reactors**

See Figure 5-9.

Storage of Power for Utilization Equipment

Batteries are appropriately located in Chapter 4, the "Use" chapter. **See Figure 5-10.**

Storage batteries are used in a variety of electrical installations to provide power for utilization equipment in the event of a power loss. **Article 480 Storage Batteries** addresses the energy storage requirements of Chapter 4 and provides information and requirements for the safe installation of these battery-supplied systems. **Article 480** is not divided into separate parts.

Figure 5-10. Batteries are equipment for general use that store energy for utilization equipment.

Article 490: Equipment, Over 1000 Volts, Nominal

The specific requirements necessary for electrical equipment for general use, rated at over 1000 volts nominal, are addressed by **Article 490**. Often, distribution voltages range between 4,160 and 13,800 volts and require special installation details. The equipment must either be indoor rated or installed outside. **See Figure 5-11.**

Figure 5-11. Typically, a transformer is mounted near the high voltage switchgear. ***Article 490*** *covers details for installations over 1000 volts.*

KEY WORDS AND CLUES FOR CHAPTER 4, "USE"

The control, connection, installation, protection, or construction specifications of general utilization equipment include the following:

- Luminaires (lighting fixtures)
- Fixture wire
- Cords and cables
- Receptacles
- Cord connectors, attachment plugs
- Switchboards
- Panelboards
- Industrial control panels
- Low-voltage lighting
- Appliances
- Fixed electric space-heating equipment
- Fixed outdoor electric deicing and snow-melting equipment
- Fixed electric heating equipment for pipelines and vessels (heat trace)
- Motors, motor circuits, and controllers
- Air-conditioning and refrigeration equipment
- Generators
- Transformers and transformer vaults
- Phase converters
- Capacitors
- Resistors and reactors
- Storage batteries
- Electrical equipment, over 1000 volts

Summary

In accordance with **Section 90.3**, *NEC* Chapter 4 applies generally to all electrical installations. Using the *Codeology* method, this chapter is called the "Use" chapter due to the scope of its coverage. All of the material covered pertains to "electrical equipment for general use." The scope of this chapter is directed at equipment that uses electrical energy and associated equipment necessary for safe utilization.

Chapter 4 of the *NEC* is titled **Equipment for General Use,** from which its scope can be described as "Information and Rules on Equipment for General Use in Electrical Installations." Chapter 4 covers the entire electrical distribution system from the service point (connection to the utility) to the last receptacle or other outlet in the electrical system for electrical equipment. Wiring methods and wiring materials are not covered here. They are found in Chapter 3.

Chapter 4, along with Chapters 1, 2, and 3, build the foundation or backbone of all electrical installations.

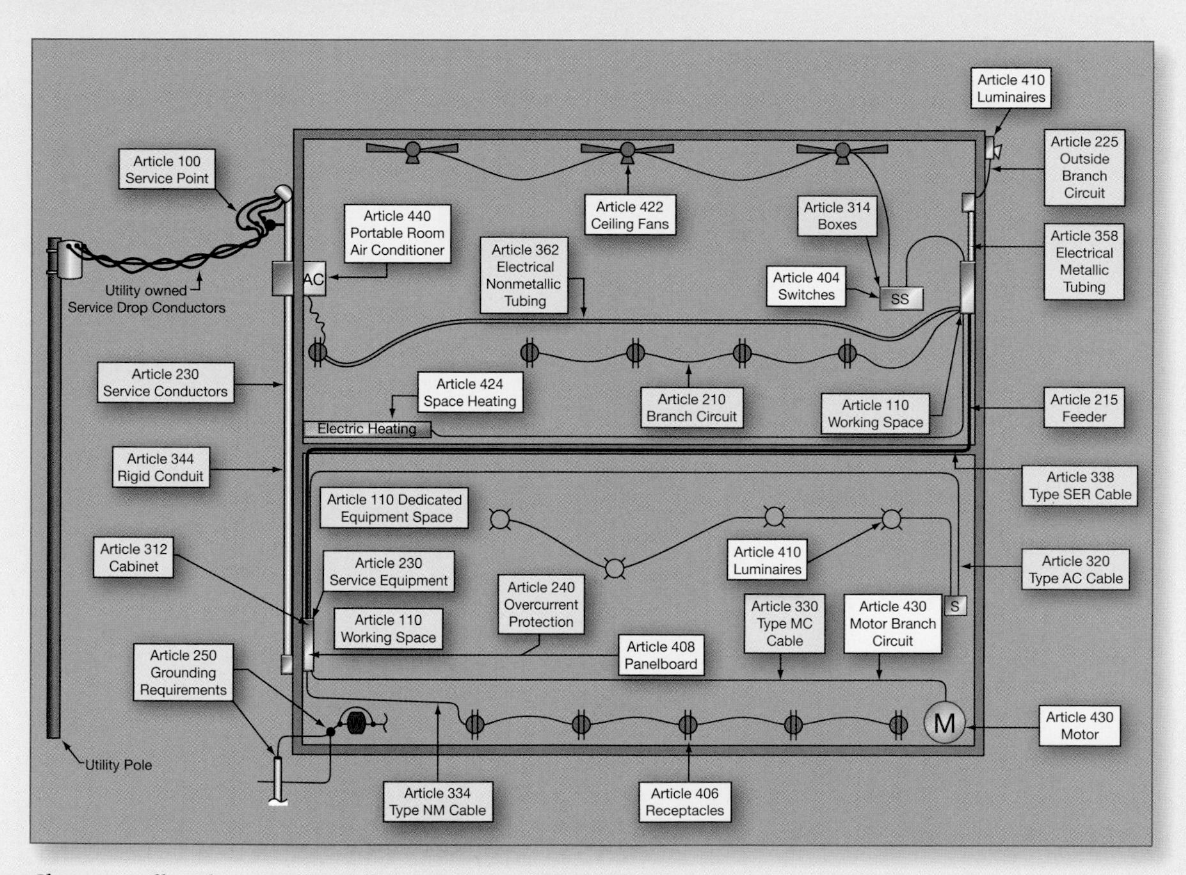

Chapter 4 will apply generally to all electrical installations.

Review Questions

1. **Chapter 4 of the *NEC* applies __?__.**
   a. generally to Chapters 5, 6, and 7
   b. to all electrical installations
   c. to only the first four chapters
   d. only to the information contained in Chapter 4
2. **Chapter 4 is subdivided into __?__ articles.**
   a. 3
   b. 22
   c. 45
   d. 46
3. **The Scope of Chapter 4 of the *NEC* is dedicated to equipment for general use to facilitate the __?__ of electrical energy.**
   a. role
   b. use
   c. none of the above
   d. all of the above
4. **Which of the following articles in Chapter 4 addresses the installation of track lighting?**
   a. **Article 400**
   b. **Article 402**
   c. **Article 410**
   d. **Article 411**
5. **Chapter 4 of the *NEC* addresses equipment for general use, which would include motors. Which section applies to sizing motor overload protection?**
   a. **430.6**
   b. **430.22**
   c. **430.32**
   d. **430.50**
6. **Which Section applies to resistance heating elements?**
   a. **424.6**
   b. **426.20**
   c. **426.30**
   d. **426.33**
7. **Which articles in Chapter 4 cover requirements for control of utilization equipment?**
   a. **Articles 440, 445, and 480**
   b. **Articles 445, 450, and 480**
   c. **Articles 445, 470, and 480**
   d. **Articles 455, 450, and 480**
8. **How many parts is Article 490 subdivided?**
   a. 1 part
   b. 3 parts
   c. 5 parts
   d. 6 parts
9. **Which part of Article 440 applies to the installation of portable room air conditioners?**
   a. **Part I**
   b. **Part II**
   c. **Part III**
   d. None of the above
10. **Which article in Chapter 4 covers requirements for generation of power utilization equipment?**
    a. **Article 445**
    b. **Article 450**
    c. **Article 470**
    d. None of the above

Chapters 5, 6, and 7 of the *NEC*, "Special"

Introduction

In accordance with **Section 90.3** of the *NEC*, Chapters 1 through 4 are general in scope and provide installation requirements for an entire electrical system. Chapters 5, 6, and 7 are the applications of special requirements, supplements, or modifications to the first seven chapters. All of the requirements in these "Special" chapters typically modify the basic rules or apply supplemental requirements to address special needs. The "Special" chapters contain 70 articles, making it necessary to become familiar with the different types of special occupancies, equipment, and conditions.

Objectives

» Associate the *Codeology* title for *NEC* Chapter 5 as "Special Occupancies," Chapter 6 as "Special Equipment," and Chapter 7 as "Special Conditions."

» Identify the special type of information and requirements contained in Chapters 5, 6, and 7 for supplementing and/or modifying the requirements of Chapters 1 through 7.

» Recognize Chapters 5, 6, and 7 as the 500-, 600-, and 700-series.

» Recognize, recall, and become familiar with articles contained in Chapters 5, 6, and 7.

Chapter 6

Table of Contents

STRUCTURE OF THE "SPECIAL" CHAPTERS

The need to address modifications and supplemental requirements covered by the "Special" chapters must be identified before any work is started on an electrical installation. Each stage of an electrical installation is affected when a special situation is encountered. Therefore, the special requirements of Chapters 5, 6, and 7 must be considered in each step of the installation to prevent serious misapplication of the *NEC*.

Imagine that a new hospital is to be constructed, and the electrical installation is being studied. A hospital is considered a special occupancy because of the specific requirements for uninterrupted power, protection of patients, electrical wiring and many other special needs. The *NEC* covers the special electrical installation requirements of a hospital in **Article 517 Health Care Facilities. See Figure 6-1.**

PLANNING STAGES MODIFIED AND/OR SUPPLEMENTED BY CHAPTERS 5, 6, AND 7

Continuing the example of planning for new hospital construction, the requirements of **Chapter 5 Special Occupancies** will affect the Plan stage of the hospital. For example, **Article 517** is intended as an electrical installation article and modifies and/or supplements some rules in the "*Plan*" stage of the electrical installation. Special requirements for the installation of branch circuits, grounding and bonding, and receptacle locations for patient care spaces move beyond requirements in other buildings. These special provisions require specific wiring methods to the patient care spaces, thereby affecting the "*Build*" stage of this installation.

The "Use" stage of this installation will be modified and supplemented by requirements, including those for therapeutic pools and tubs, in **Chapter 6 Special Equipment.** All stages of the installation for the life safety system will be further modified and supplemented by **Chapter 7 Special Conditions** when essential electrical systems are installed to meet critical care, life safety, and equipment installation requirements.

Figure 6-1 Hospitals Are Special Occupancies

Deaconess Riley Hospital for Children

Figure 6-1. Chapter 5 covers special occupancies such as hospitals.

NEC CHAPTER 5, "SPECIAL OCCUPANCIES"

The *Codeology* title for *NEC* Chapter 5 is "Special Occupancies." All stages of an electrical installation will be modified or supplemented when special occupancies are involved. The 28 articles of Chapter 5 can be grouped as follows:

Hazardous Locations

Article 500 Hazardous (Classified) Locations, Classes I, II, and III, Divisions 1 and 2
Article 501 Class I Locations
Article 502 Class II Locations
Article 503 Class III Locations
Article 504 Intrinsically Safe Systems

Hazardous Locations – Zone Classification Systems

Article 505 Zone 0, 1, and 2 Locations
Article 506 Zone 20, 21, and 22 Locations for Combustible Dusts or Ignitable Fibers/Flyings

Article 510 Hazardous (Classified) Locations – Specific
Article 511 Commercial Garages, Repair and Storage
Article 513 Aircraft Hangars
Article 514 Motor Fuel Dispensing Facilities
Article 515 Bulk Storage Plants
Article 516 Spray Application, Dipping and Coating Processes

Health Care Facilities
Article 517 Health Care Facilities

Assembly Occupancies for 100 or More Persons
Article 518 Assembly Occupancies

Entertainment Venues
Article 520 Theatres, Audience Areas of Motion Picture and Television Studios, Performance Areas, and Similar Locations
Article 522 Control Systems for Permanent Amusement Attractions
Article 525 Carnivals, Circuses, Fairs, and Similar Events
Article 530 Motion Picture and Television Studios and Similar Locations
Article 540 Motion Picture Projection Rooms

Agricultural Buildings,
Article 547 Agricultural Buildings

Manufactured Buildings, Dwellings, and Recreational Structures
Article 545 Manufactured Buildings
Article 550 Mobile Homes, Manufactured Homes, and Mobile Home Parks
Article 551 Recreational Vehicles and Recreational Vehicle Parks
Article 552 Park Trailers

Structures on or Adjacent to Bodies of Water
Article 553 Floating Buildings
Article 555 Marinas and Boatyards

Temporary Installations
Article 590 Temporary Installations

Hazardous Locations

Article 500 through **Article 516** contain modifications and supplemental requirements for occupancies that contain, process, manufacture, or store materials that could cause a fire or explosion due to flammable gases or vapors, flammable liquids, combustible dust(s) or ignitable fibers and flyings. **See Figure 6-2.**

An important part of a new *Code* user's experience is becoming familiar with hazardous locations. Users will often think of industrial locations as the only hazardous locations, but in reality, the general public typically visits one of the most common hazardous locations at least once a week. Portions of gas stations used to refuel vehicles fall under the classification of hazardous locations. Only the area where the gas is being dispensed is considered hazardous, however; the convenience store associated with the typical gas station is not considered a hazardous location and is covered by Chapters 1 through 4 of the *NEC*. **See Figure 6-3.**

Figure 6-2 Industrial Chemical Processing Plant

Figure 6-2. Hazardous locations that process ignitable materials, liquids and vapors are special occupancies.

Figure 6-3. Not all areas of a gas station are considered hazardous. The inside of a convenience store associated with the gas station is not considered a hazardous location or a special occupancy.

Article 500 sets the stage for the application of **Articles 501, 502, 503,** and **504. Section 500.2** contains definitions that apply to all hazardous location articles (**Article 500** through **Article 516**), except for the two zone classification articles (**Article 505** and **Article 506**), which provide an alternative to the Class I, II, and III systems. **Article 500** provides the basic information and requirements for the application of Class I, II, and III systems for hazardous locations. The following key provisions of **Article 500** apply to all hazardous location articles except for zone system **Articles 505** and **506. 500.5** explains how locations are classified. The informational notes found in each part of **500.5** give examples of occupancies where the user would find those classified locations.

Article 500 Hazardous (Classified) Locations, Classes I, II, and III, Divisions 1 and 2

Electrical installations in classification areas require electrical materials and equipment designed to prevent sparks from occurring around flammable vapors, liquids, gases, and dusts/fibers. These items are not used in basic commercial and industrial locations and may be unfamiliar to the installer. Items such as hazardous-rated push button switches, receptacles, and light fixtures must be installed in classified areas. **See Figure 6-4.**

Not only does electrical equipment that produces sparks need to be hazardous rated, so does the conduit system. The conduit system has to prevent ignitable vapors, liquids, gases, and dust/fibers from entering the raceway and contacting the equipment, creating sparks. Junction boxes, flexible connections, and 90-degree pull boxes must all be hazardous rated. **See Figure 6-5.**

500.5(A) Classifications of Locations

The classification of hazardous locations is based upon two factors:

1. The properties of the flammable vapors, liquids, gases or flammable liquid-produced vapors, or combustible dusts or fibers that may be present

For additional information, visit qr.njatcdb.org Item #1072

Figure 6-4. Equipment that creates sparks when utilized must be designed so as not to ignite flammable vapors, liquids, gases, and dusts/fibers.

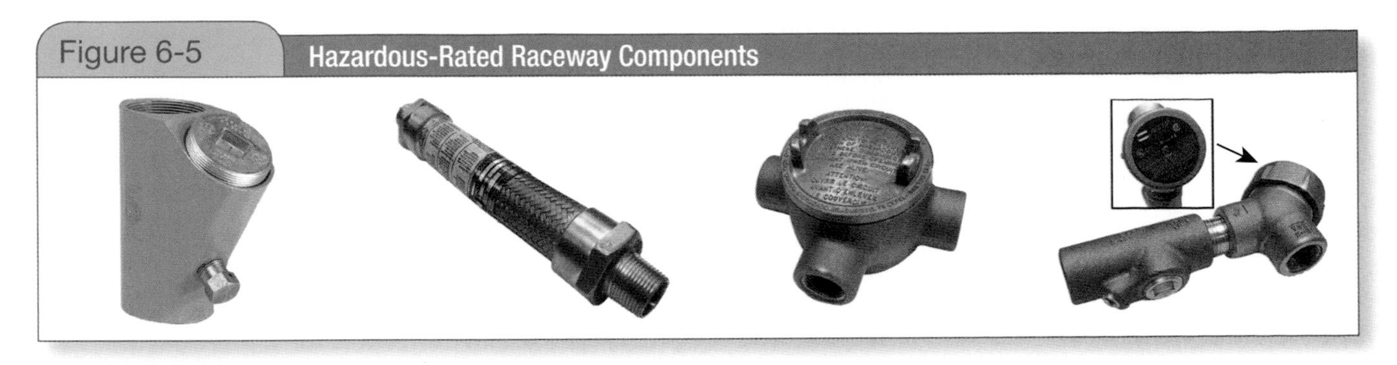

Figure 6-5. Equipment that creates sparks when utilized must be connected with a raceway system that is installed for hazardous areas.

2. The likelihood that a flammable or combustible concentration or quantity is present

500.5(B) Class I Locations

Class I locations are those in which flammable gases or flammable liquid-produced vapors are or may be present in the air in quantities sufficient to produce explosive or ignitable mixtures.

(1) Class I, Division 1

- Locations in which ignitable concentrations exist under normal operations
- Locations in which ignitable concentrations may exist due to repair, maintenance, or leaks
- Locations in which ignitable concentrations exist due to processes, breakdown, or faulty equipment

(2) Class I, Division 2

- Locations in which volatiles are handled, processed, or used in closed containers
- Locations in which positive ventilation prevents accumulation of gases/vapors
- Areas adjacent to Class I, Division 1 locations

500.5(C) Class II Locations

Class II locations are those that are hazardous because of the presence of combustible dust. An example of a Class II location is a grain storage bin that produces combustible dust during the movement of the grain. **See Figure 6-6.**

Other Class II hazardous location examples are flour and feed mills; producers of plastics, medicines, and fireworks; producers of starch or candies; spice-grinding plants; sugar plants; and cocoa plants.

(1) Class II, Division 1

- Locations in which ignitable concentrations of combustible dust exist under normal operations
- Where mechanical or machinery failure, repair, maintenance, or leaks could create ignitable concentrations of dust
- Locations in which metal dusts exist, including, but not limited to, aluminum and magnesium

For additional information, visit qr.njatcdb.org Item #1073

Figure 6-6
Storage of Grain Indicates a Hazardous Location

Figure 6-6. Across the heartland of America, farmers utilize grain storage bins that must be hazardous rated. It makes no difference whether it is a couple of storage bins or a large distribution center; they are all considered hazardous.

(2) Class II, Division 2

- Locations in which ignitable concentrations of combustible dust may exist due to abnormal operations
- Locations in which combustible dust is present, but not in sufficient amounts, unless a malfunction of equipment or process occurs
- Locations in which accumulating dust interferes with heat dissipation and/or could be ignited through equipment failure

500.5(D) Class III Locations

Class III locations are those that are hazardous because of the presence of easily ignitable fibers or flyings, but in which such fibers or flyings are not likely to be in suspension in the air in quantities sufficient to produce ignitable mixtures.

Class III, Division 1

- Locations in which easily ignitable fibers or materials producing combustible flyings are handled, manufactured or used

Class III, Division 2

- Locations in which easily ignitable fibers are stored or handled other than in the process of manufacture

Articles 501, 502 and 503 cover Class 1, 2 and 3 Hazardous Locations. These articles have their own parallel numbering system similar to what is found in Chapter 3 of the *NEC* with raceways. All section topics do not carry over from **Article 501 to 502 and 503.** Some hazardous locations have hazards unique to their class. This number aids the user in finding the relevant section quickly. **See Figure 6-7.**

Specific Class I, II, and III locations

Article 510 sets the stage for the application of **Article 511** through **Article 517.** Note that **Article 510** references **Article 517** for health care facilities because of the explosive anesthetic gases or vapors that may be used in inhalation anesthetization. In the United States, flammable anesthetic gases such as cyclopropane are not allowed for use in health care facilities. However, some other countries still allow their use. **Article 517** is not always grouped with the specific Class I, II, and III locations due to the many other requirements that make **Article 517** special.

Health care facilities occupy buildings or portions of buildings in which they provide medical, dental, psychiatric, nursing, obstetrical, or surgical care. **See Figure 6-8.** Health care facilities include,

Figure 6-7 **Parallel Numbering Sections in Articles 501, 502, and 503**

Article 501 Class 1 Locations	Article 502 Class 2 Locations	Article 503 Class 3 Locations
501.10 Wiring Methods	502.10 Wiring Methods	503.10 Wiring Methods
501.15 Sealing	502.15 Sealing	503.15 Sealing
501.30 Grounding and Bonding	502.30 Grounding and Bonding	503.30 Grounding and Bonding
501.100 Transformers, Capacitors	502.100 Transformers, Capacitors	503.100 Transformers, Capacitors
501.125 Motors & Generators	502.125 Motors & Generators	503.125 Motors & Generators
501.130 Luminaires	502.130 Luminaires	503.130 Luminaires
501.150 Signaling, Alarm, Comm.	502.150 Signaling, Alarm, Comm.	503.150 Signaling, Alarm, Comm.

Figure 6-7. *Where possible, similar subjects with the same purposes (for example,* ***501.30 Grounding and Bonding****) are used in section numbers, and part numbers for the same purposes within* ***Articles** 501, 502* *and* *503.*

Figure 6-8 Health Care

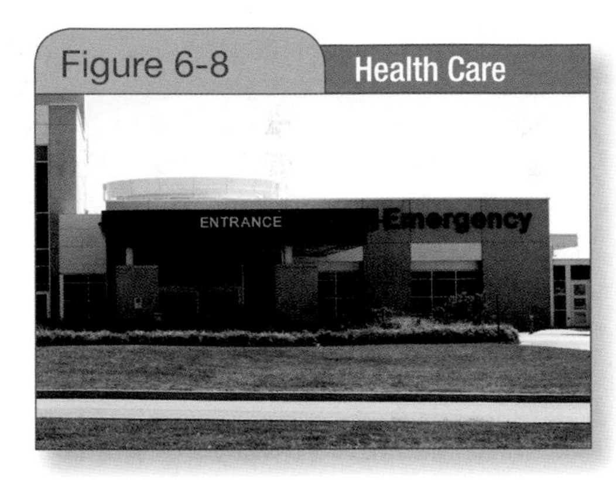

Figure 6-8. Chapter 5 of the NEC *details the special installation requirements for facilities such as hospitals (***Article 517***).*

but are not limited to, hospitals, nursing homes, limited care facilities, clinics, medical and dental offices, and ambulatory care centers, whether permanent or movable. The definition of health care facilities in **Section 517.2** outlines locations covered by this article. *NFPA 99: Health Care Facilities Code* provides many design, installation, and maintenance rules required for health care installations.

Other Chapter 5 Articles

Assembly Occupancies – **Article 518.** Assembly occupancies have additional wiring requirements and grounding requirements, See **518.4**. For examples of assembly occupancies, **See Figure 6-9.**

Examples of assembly occupancies include, but are not limited to, the following:

- Armories
- Assembly Halls
- Auditoriums
- Bowling Lanes
- Club Rooms
- Conference Rooms
- Courtrooms
- Dance Halls
- Dining/Drinking Facilities
- Exhibition Halls
- Gymnasiums
- Mortuary Chapels
- Multipurpose Rooms
- Museums
- Places of Awaiting Transportation
- Places of Religious Worship

Figure 6-9 Assembly Location – Worship Center

Figure 6-9. Worship centers are one of the most common examples of an assembly occupancy.

- Pool Rooms
- Restaurants
- Skating Rinks

Assembly occupancies also include:

- Entertainment Venues. **See Figure 6-10.**
- **Article 547 Agricultural Buildings.** In agricultural buildings where livestock is confined, such as a dairy, an equipotential plane must be established per **547.10**. All of the metal paths that may become energized and ground where the livestock are present are kept at the same potential. Small voltage potentials can harm livestock and in a dairy would have a negative effect on production.
- Manufactured Buildings, Dwellings, and Recreational Structures.

Figure 6-10 Assembly Location – Movie Theater

Figure 6-10. Movie theaters are entertainment venues typically with more than 100 occupants, therefore falling under the special conditions of Chapter 5.

Figure 6-11 Electrical Panels Near Bodies of Water

Figure 6-11. Articles 553 and 555 detail the requirements for electrical installations near bodies of water.

Figure 6-12 Temporary Power

Figure 6-12. Article 590 allows for temporary power installations for construction and other considerations.

- Structures on or Adjacent to Bodies of Water. **See Figure 6-11.**
- Temporary Installations. **See Figure 6-12.**

NEC CHAPTER 6, "SPECIAL EQUIPMENT"

The *Codeology* title for Chapter 6 is "Special Equipment." While Chapter 4 of the *NEC* is titled **Equipment for General Use** and covers the basics, Chapter 6 covers special equipment and modifies or supplements the first seven chapters with 27 articles to meet the special needs of the equipment covered in the chapter.

- **Article 600 Electric Signs and Outline Lighting**
- **Article 604 Manufactured Wiring Systems**
- **Article 605 Office Furnishings (Consisting of Lighting Accessories and Wired Partitions)**
- **Article 610 Cranes and Hoists**
- **Article 620 Elevators, Dumbwaiters, Escalators, Moving Walks, Platform Lifts, and Stairway Chairlifts**
- **Article 625 Electric Vehicle Charging System**
- **Article 626 Electrified Truck Parking Spaces**
- **Article 630 Electric Welders**
- **Article 640 Audio Signal Processing, Amplification, and Reproduction Equipment**
- **Article 645 Information Technology Equipment**
- **Article 646 Modular Data Centers**
- **Article 647 Sensitive Electronic Equipment**
- **Article 650 Pipe Organs**
- **Article 660 X-Ray Equipment**
- **Article 665 Induction and Dielectric Heating Equipment.**
- **Article 668 Electrolytic Cells**
- **Article 669 Electroplating**
- **Article 670 Industrial Machinery**
- **Article 675 Electrically Driven or Controlled Irrigation Machines**
- **Article 680 Swimming Pools, Fountains, and Similar Installations**
- **Article 682 Natural and Artificially Made Bodies of Water**
- **Article 685 Integrated Electrical Systems**
- **Article 690 Solar Photovoltaic (PV) Systems. See Figure 6-13.**
- **Article 691 Large-Scale Photovoltaic (PV) Electric Supply Stations**
- **Article 692 Fuel Cell Systems. See Figure 6-14.**

Figure 6-13 Solar PV Cells

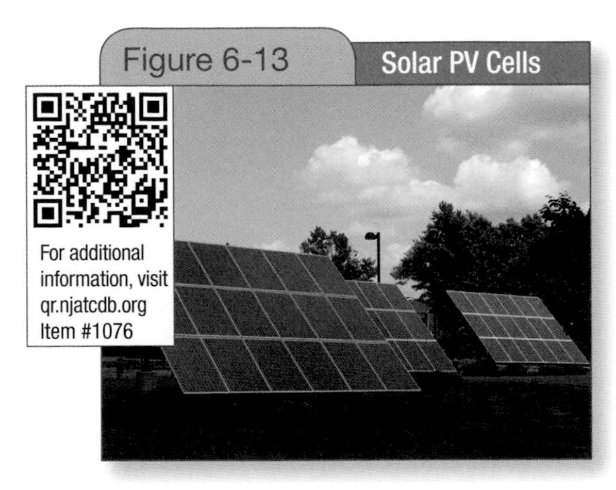

For additional information, visit qr.njatcdb.org Item #1076

Figure 6-13. Solar photovoltaic systems are special equipment addressed in Chapter 6.

Figure 6-14 Residential Fuel Cell

Figure 6-14. Fuel cell systems are special equipment addressed in Chapter 6 (Article 692).

Figure 6-15 Fire Pump and Motor

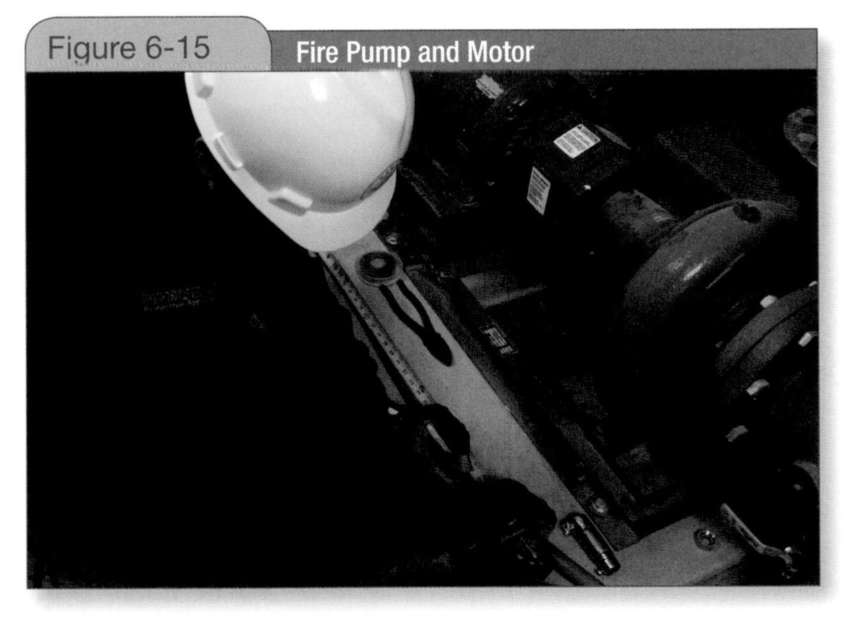

Figure 6-15. Fire pumps are very critical to the safety of building occupants. Electrical installation requirements are covered in **Article 695**.

- **Article 694 Small Wind Electric Systems**
- **Article 695 Fire Pumps. See Figure 6-15.**

Article 680 Swimming Pools, Fountains and Similar Installations

Swimming pools present their own hazards when combined with the use of electricity which must be prevented with the proper application of the *NEC*. **Article 680** is divided into 8 parts. The user must pay attention to each part to properly apply the *NEC*. For example, **Part III** of **Article 680** has requirements or allowances that are unique to Storable Pools and does not apply to Permanent Pools in **Part II**. **Part I** of **Article 680** applies to all parts of the article unless otherwise modified by a particular part.

In **Part I**, **Article 680.9** discusses the clearance requirements for conductors when they are installed over a pool to prevent people from accessing the conductors and potentially energizing the pool or surrounding equipment. Note that this is in **Part 1** and therefore applies to the rest of the article unless amended.

The use of electrical equipment such as radios, chargers for electronics, or other devices next to a pool can be dangerous. **Article 680.22** requires a receptacle at a permanently installed pool to allow the use of a radio or other equipment while enjoying the pool. The receptacle must be at least six feet from the edge of the water, to prevent the equipment from being put into the pool while energized. This receptacle must also be within 20 feet of the inside edge of the pool, eliminating the possibility that someone may use an extension cord to put the equipment closer for their convenience, adding additional hazard. **Article 680.22** also requires receptacles within 20 feet of the inside walls of the pool to be GFCI protected. Low voltage applications are not exempt from requirements, **Article 680.22(D)** requires other outlets, which would include telephone, data, or fire alarm, to be at least ten feet away from the inside walls of the pool.

To further reduce hazards, **Article 680.26** has special bonding requirements of the equipment around the pool, the pool shell, and the surface around the pool, and measures must be taken to assure the water in the pool is bonded. See **680.26 (B)1 through (B)7** for more information. These requirements limit the voltage potential and put the area of the pool at the same voltage, reducing electrical currents due to stray voltages.

Other Chapter 6 Articles

Article 680 demonstrates it is important to be familiar with the articles that make up Chapter 6. An Electrical Worker must know that **Article 630** exists to apply the requirements when installing the circuits necessary for a group of welders. To do otherwise would result in costly delays. Users must familiarize themselves with all the articles in Chapter 6.

NEC CHAPTER 7, "SPECIAL CONDITIONS"

The *Codeology* title for Chapter 7 is "Special Conditions" and covers conditions required to meet the special needs of different types of occupancies and systems. For example, when an alternate power source is required for emergency or legally required standby, it must be installed according to Chapters 1 through 4 and modified and/or supplemented in accordance with the requirements of **Chapter 7 Special Conditions.**

- **Article 700 Emergency Systems – See Figure 6-16.**

Figure 6-16 Emergency System

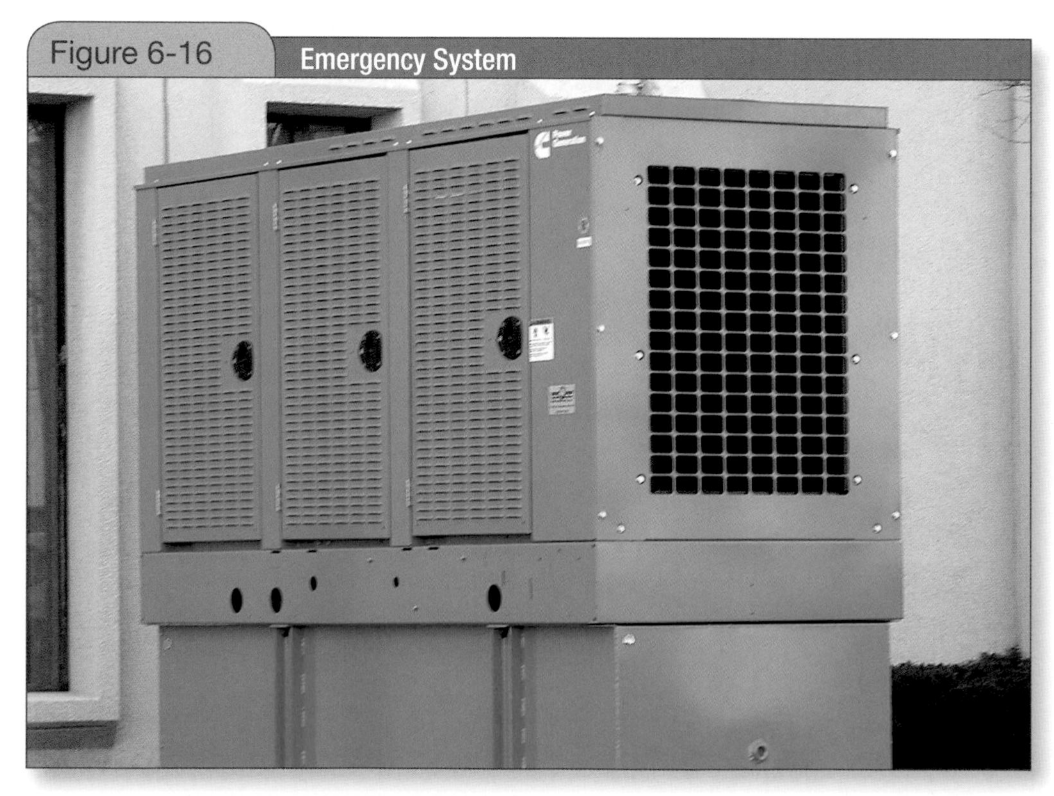

Figure 6-16. *Emergency systems use a standby generator as the emergency source in many installations and represent special conditions addressed in Chapter 7.*

- **Article 701 Legally Required Standby Systems**
- **Article 702 Optional Standby Systems**
- **Article 705 Interconnected Electric Power Production Sources**
- **Article 706 Energy Storage Systems**
- **Article 708 Critical Operations Power Systems (COPS)**
- **Article 710 Stand Alone Systems**
- **Article 712 Direct Current Microgrids**
- **Article 720 Circuits and Equipment Operating at Less Than 50 Volts**
- **Article 725 Class 1, Class 2, and Class 3 Remote-Control, Signaling, and Power-Limited Circuits**
- **Article 727 Instrumentation Tray Cable: Type ITC**
- **Article 728 Fire-Resistive Cable Systems**
- **Article 750 Energy Management Systems**
- **Article 760 Fire Alarm Systems – See Figure 6-17.**
- **Article 770 Optical Fiber Cables and Raceways – See Figure 6-18.**

Emergency and Standby Power Systems

Articles 700, **701** and **702** cover emergency systems. They are presented in order of importance or criticality.

Article 700 covers emergency systems that are directly related to and essential to life safety. These systems are commonly found in hospitals and other healthcare facilities, where a power interruption would have a direct impact on human life. See **700.2, Informational Note**.

The systems in **Article 701** are those systems that are legally required by the local jurisdiction, or due to the occupancy and use of the facility require a standby power system. **Article 701** covers systems that provide power to lighting or other selected loads that without power, would hamper rescue operations or create an unsafe condition. See the **Informational Note** on **701.2**.

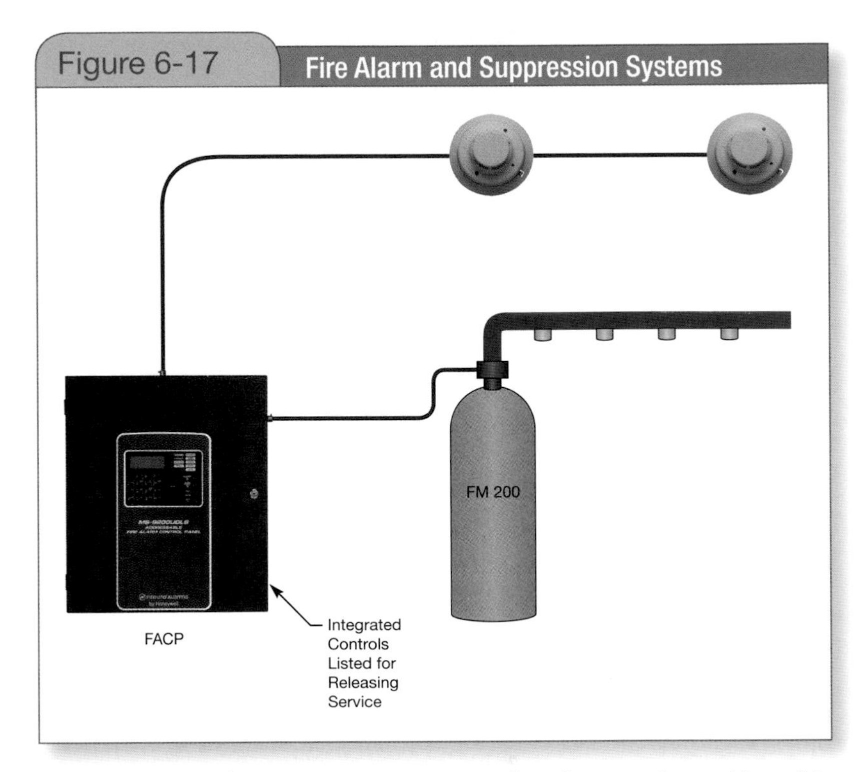

Figure 6-17. Fire alarm systems represent special conditions and are addressed in Chapter 7. Further requirements are also listed in NFPA 72: National Fire Alarm and Signaling Code.

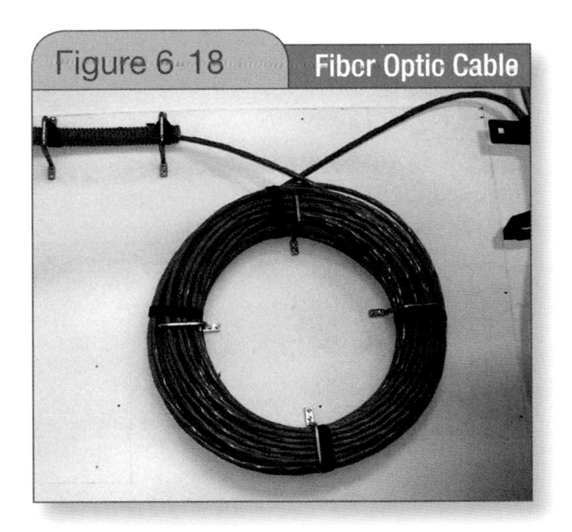

Figure 6-18. When fiber optic cable and raceways are installed, they represent a special condition addressed in Chapter 7.

Article 702 is optional standby systems. These systems are typically installed at the request of the owner to prevent loss of product or inventory or to provide mission continuity of a business through a power outage. A restaurant or grocery store may want backup power for the refrigeration to prevent spoilage, or an Internet service provider would want backup power to provide continuous service to their customers.

A quick look at these articles will show how the criticality is reflected through the requirements of each article. **Section 700.10(B) Wiring Methods** states that the wiring from the emergency power shall generally not occupy the same raceway or enclosure as the normal power. In **702.10**, the two systems may share the same enclosure and raceway, and the same is true for **703.10**.

When normal power is lost, **Articles 701** and **702** have different time allowances for emergency power to be operational and online. Ten seconds is required on an emergency system in **Article 701** and a standby power system in **Article 702** may take up to 60 seconds. There is no time requirement in **Article 703 Optional Standby Systems**.

Class 1, 2 and 3 Remote Control, Signaling, and Power Limited Circuits

The majority of **Article 725** covers power limited circuits. Many electricians see this article as the "low voltage" article, but it is important to note that this article also covers non-power limited class 1 circuits, which according to **725.41(B)** can be up to 600 volts. A typical example of a non-power limited class 1 circuit would be a motor control circuit.

Articles in Chapter 7 modify Chapters 1 through 7. **725.3** states that only those sections of **Article 300** that are mentioned in **725** apply to Class 1, Class 2 and Class 3 circuits. There are many articles in the 300 section that apply to general wiring methods that do not apply to Class 1, 2 and 3 circuits.

Class 2 and Class 3 circuits generally may not share a raceway or enclosure with a power or lighting circuit. This prevents electrical interference and, more importantly, prevents energizing the Class 2 or Class 3 circuits with a power or lighting circuit in the case of a fault. See **725.136(A)** and **725.136(B) through (G)** for additional allowances. For example, **725.136(D)** allows power and lighting to share the same enclosure in a limited instance.

This is the case with a security panel that has 120-volt power and the class 3 security circuits in the same enclosure that serve the same system.

Other Chapter 7 Articles

It is important to be familiar with articles in Chapter 7. In the last two *Code* cycles alone, many new articles such as **Articles 706, 710, 712, 728,** and **750** have been added. Technology and innovative ways to provide and control power are continuously coming into the electrical industry. Perhaps no other chapter in the *NEC* illustrates this more than Chapter 7.

Summary

Chapters 5, 6, and 7 are known as the "Special Chapters" because they contain modifications, supplemental information, and rules that complement or complete the basic foundation of Chapters 1 through 4. These special chapters contain 70 articles that the *Code* user must become familiar with for proper application. When applying the rules of the *NEC* to any electrical installation, it is imperative that the user recognizes and refers to the special chapters whenever special occupancies, special equipment, or special conditions are part of an installation.

Review Questions

1. **Chapters 5, 6, and 7 of the *NEC* __?__**
   a. apply generally to all electrical installations.
   b. apply only to information in Chapters 5, 6, and 7
   c. supplement or modify Chapters 1 through 7
   d. none of the above
2. **Chapter 6 of the *NEC* has __?__ articles.**
   a. 24
   b. 25
   c. 26
   d. 27
3. **The scope of Chapter 5 of the *NEC* is dedicated to Special __?__**
   a. Conditions
   b. Equipment
   c. Occupancies
   d. Wiring
4. **Which article and part address the installation of a temporary audio installation?**
   a. **Article 640 Part I**
   b. **Article 640 Part II**
   c. **Article 640 Part III**
   d. All of the above
5. **Name the part and article in Chapter 7 that would apply when sizing overcurrent protection in a legally required standby system.**
   a. **Article 700 Part VI**
   b. **Article 708 Part IV**
   c. **Article 701 Part IV**
   d. None of the above
6. **All of the definitions in Section 500.2 apply to Article 501.**
   a. True
   b. False
7. **Which articles in Chapter 5 address entertainment venues?**
   a. **Articles 520, 522,** and **525**
   b. **Article 520, 522,** and **555**
   c. **Articles 530** and **540**
   d. A and C
   e. B and C
8. **Article 690 has __?__ parts**
   a. 9
   b. 10
   c. 11
   d. None of the above
9. **Part III of Article 725 applies to Class 1 circuits.**
   a. True
   b. False
10. **Which special article covers requirements for generation of power with fuel cells?**
    a. **Article 680**
    b. **Article 690**
    c. **Article 692**
    d. **Article 692.8**

Chapter 8 – Communications, and Chapter 9 – Tables and Informative Annexes

Introduction

The *Codeology* title for Chapter 8 of the *NEC* is "Communications." Many would say that this chapter is an island, standing apart from the rest of the *Code*. In **90.3 *Code* Arrangement**, Chapter 8 covers communications systems. It is the only *Code* chapter that is not subject to the requirements of Chapters 1 through 7, except where requirements are specifically referenced in Chapter 8. Five articles comprise Chapter 8, the 800-series. These five articles provide the general requirements and information for all installations of communications systems.

Chapter 9 consists of twelve Tables and ten Informative Annexes. Tables are applicable only where referenced in the *NEC*; see **344.22** for an example. Annexes are for informational use only and are not enforceable parts of the *Code*. These Tables and Informative Annexes supply the *Code* user with data to calculate conduit fill, ampacity, minimum raceway bending radius, voltage drop, power source limitations, and torque tightening values. They also include product safety standards, types of construction classifications, Functional Performance Tests (FPTs), Supervisory Control and Data Acquisition (SCADA) systems applications, and Americans with Disability Act (ADA) requirements.

Chapter 7

Objectives

- Associate the *Codeology* title for *NEC* Chapter 8 as "Communications."
- Describe the type of information and requirements for communications systems covered in Chapter 8.
- Recognize Chapter 8 as the 800-series.
- Recognize, recall, and become familiar with articles contained in Chapter 8.
- Associate the *Codeology* title for *NEC* Chapter 9 as "Tables and Informative Annexes."
- Identify the specific type of information contained in the Tables and Informative Annexes of Chapter 9.
- Understand that Tables in Chapter 9 apply only if referenced elsewhere in the *NEC*.
- Recognize that Informative Annexes are included in the *NEC* for informational purposes only.
- Distinguish key words and clues for locating references to Chapter 9, the "Table and Informative Annexes" chapter.
- Recognize, recall, and become familiar with Tables and Informative Annexes contained in Chapter 9.

Table of Contents

NEC CHAPTER 8, "COMMUNICATIONS SYSTEMS"

Chapters 1 through 7 of the *NEC* do not apply to any of the five articles of Chapter 8 unless specific reference is made within these articles to another area of the *NEC*. For example, all five articles of Chapter 8 specifically reference **Article 100 Definitions,** and each recognizes that all of the definitions in **Article 100** are applicable within that article. Four of the articles (**800**, **820**, **830**, and **840**) include additional definitions which are applicable only within their article. Review each definition given in these sections:

- **Article 800 - Section 800.2** recognizes all of **Article 100** and adds ten definitions which apply only within **Article 800**.
- **Article 810 - Section 810.2** recognizes all of **Article 100** and adds no additional definitions.
- **Article 820 - Section 820.2** recognizes all of **Article 100** and adds four definitions that apply only within **Article 820**.
- **Article 830 - Section 830.2** recognizes all of **Article 100** and adds seven definitions that apply only within **Article 830**.
- **Article 840 - Section 840.2** recognizes all of **Article 100** and adds three definitions which apply only within **Article 840**.

For example, **800.2 Definitions** lists several additional definitions which are somewhat unique to the other chapters of the *NEC*. It is important for the Electrical Worker to understand these terms.

Review these definitions:

- Abandoned Communications Cable
- Communications Circuit Integrity (CI) Cable
- Point of Entrance
- Premises

Each of the five articles of Chapter 8 dedicates a section for referencing other articles. This parallel numbering system is in the third section (XXX.3) of the article (for example, **Section 800.3**). Among the articles recognized outside of Chapter 8 are the requirements of *NEC* Chapter 5 where hazardous locations are encountered. Also, see **300.22(C)** for permitted wiring methods in spaces used for environmental air. These examples are not all-inclusive; multiple references to other requirements in Chapters 1 through 7 appear in the five articles of Chapter 8.

While the rest of the *NEC* does not apply to Chapter 8 unless specifically referenced, all five articles of Chapter 8 must be enforced once the *Code* is adopted. The *NEC* covers signaling and communications conductors, equipment, and raceways and is intended to give governmental agencies legal jurisdiction over electrical installations including signaling and communications systems, per **Section 90.4**. Articles of Chapter 8 cover specific types, methods, conductors, and equipment for communications systems. **See Figure 7-1.**

Figure 7-1 Layout of *NEC* Chapter 8

***NEC*® Title:**	**Communications Systems**
***Codeology* Title:**	**Communications**
Chapter Scope:	***Communications Systems* Only**

Article	Article Title
800	Communications Circuits
810	Radio and Television Equipment
820	Community Antenna Television and Radio Distribution Systems
830	Network-Powered Broadband Communications Systems
840	Premises-Powered Broadband Communications Systems

Figure 7-1. Chapter 8 contains five articles which, for the most part, stand alone from the rest of the chapters in the NEC.

Article 800 Communications Circuits

Article 800 is divided into six parts. See the Table of Contents for a list of the Parts in this Article. **See Figure 7-2.**

The six parts of **Article 800**:

I. **General**
II. **Wires and Cables Outside and Entering Buildings**

III. Protection
IV. Grounding Methods
V. Installation Methods Within Buildings
VI. Listing Requirements

Grounding

An important part of Chapter 8 that is often taken lightly by the Electrical Worker is grounding of communications systems. **Part IV** of Chapter 8 covers grounding and bonding methods required for communications systems. Not only does proper grounding reduce safety hazards, but most often ensures proper system performance. For instance, static noise heard on a telephone line is typically the result of poor grounding in the system. If the same idea is applied to a communications cable transmitting data, the performance and accuracy of the data will suffer. Proper installation of the bonding conductor or grounding electrode conductor (GEC) is covered thoroughly throughout Chapter 8.

Raceways

Communications conductors can be installed in various raceways as described in a portion of **Part V** of Chapter 8. **800.100(A)-(L)** lists uses of raceways in various environmental spaces.

Unlike power conductors, which can be installed together in the same raceway, grouping lighting and power conductors with communications cables in the same raceway will cause performance issues for the communications system and in most cases is a *Code* violation.

800.133 Installation of Communications Wires, Cables, and Equipment covers the separation of communications cables from power cables. As an installer, this section must be understood in detail. Innerduct can be used to separate different communications cables within the same conduit or raceway. When a larger conduit is installed, it allows for additional cables of different systems to be added in the future. **See Figure 7-3.**

Figure 7-2 Telephone M66 Blocks

Figure 7-2. Chapter 8 covers the installation requirements for communications cables such as telephone systems.

Figure 7-3 Innerduct

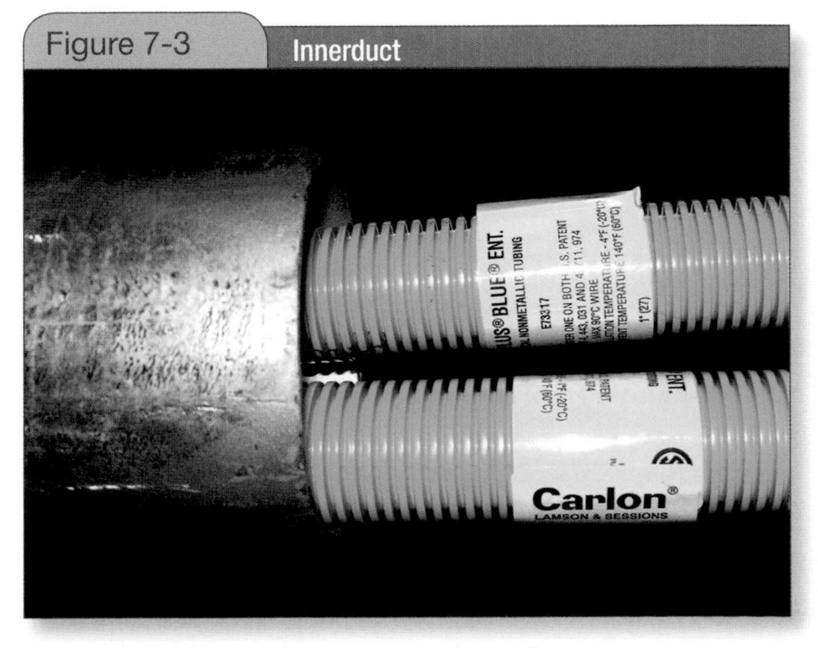

Figure 7-3. Installing innerduct in conduit allows different communications systems' cables to be separated. This also allows for future cable installations.

For additional information, visit qr.njatcdb.org Item #2511

Figure 7-4 Bridle Ring

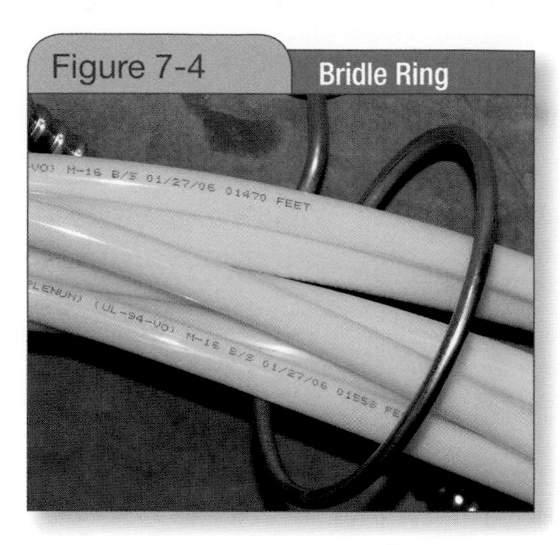

Figure 7-4. It is common to install communications cables above a lay-in ceiling with bridle rings (or other cable management methods) instead of conduit.

It is common to install communications conductors without the use of raceways above lay-in ceilings. The *NEC* requires communication cables to be installed in a neat and workmanlike manner. Cables are required to be supported by the building structure using support hardware such as hangers, straps, staples, cable ties, and so on. Communications rings are often used since they do not damage the cable and are easy to install. **See Figure 7-4.** Please note, cables cannot be supported by lay-in ceiling tiles or attached to the lay-in ceiling support wires; or, for that matter, even strapped to the conduit. Strapping cables to a conduit will affect the ability of heat to escape, therefore reducing the ability of the power conductor to carry the proper current.

Abandoned cables are defined as communication cables that are not terminated at both ends to a connector or equipment and are not identified by a tag for future use. These cables shall be removed unless identified (tagged) for future use.

Figure 7-5 Plenum Ceiling

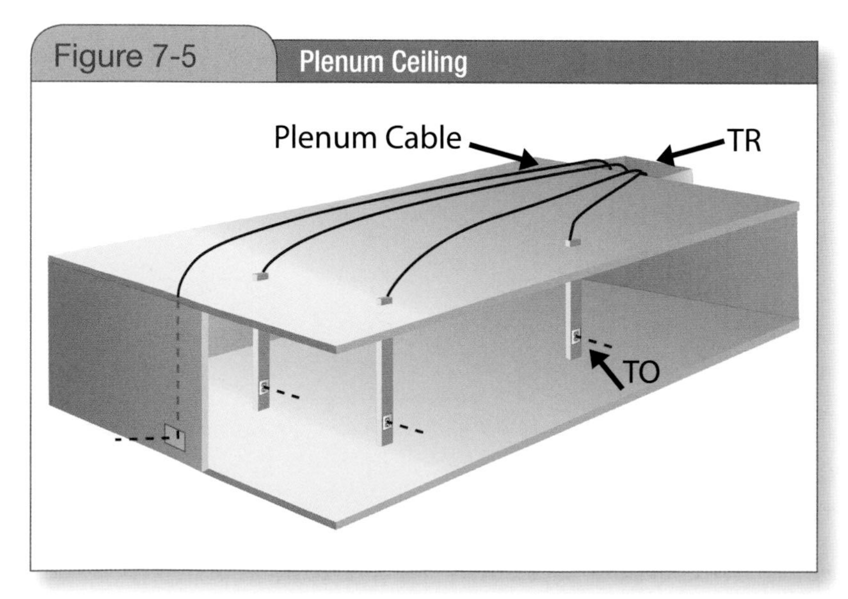

Figure 7-5. It is common to install communications cables above a lay-in ceiling. But if not installed in a raceway, the cable shall be listed for plenum installations.

Listing and Requirements

Part VI of Chapter 8 covers the required listing of a communications cable's outer jacket. Most communications systems installations do not require these cables to be installed in raceways. However, there are requirements in the *NEC* that cover installations and wiring methods for communications systems. **Section 800.154** describes the listing (rating) of the communications cable jacket in certain installations. For example, a plenum space is where there is no supply or return ductwork for the HVAC. Oftentimes the space about the lay-in ceiling is known as a plenum space. In most plenum situations, this space provides the return air path back to the HVAC equipment. Because of this situation, the jacket on the communications cables must have limited smoke characteristics in case of fire. Though it is beyond the intent of *Codeology* to explain all of the differences between CMP, CMR, CMG/CM, CMX, and CMUC rated jackets, it would be advisable that the Electrical Worker become very familiar with **Part VI** of Chapter 8 before purchasing and installing communications cables outside of raceways in plenum ceilings. **See Figure 7-5.**

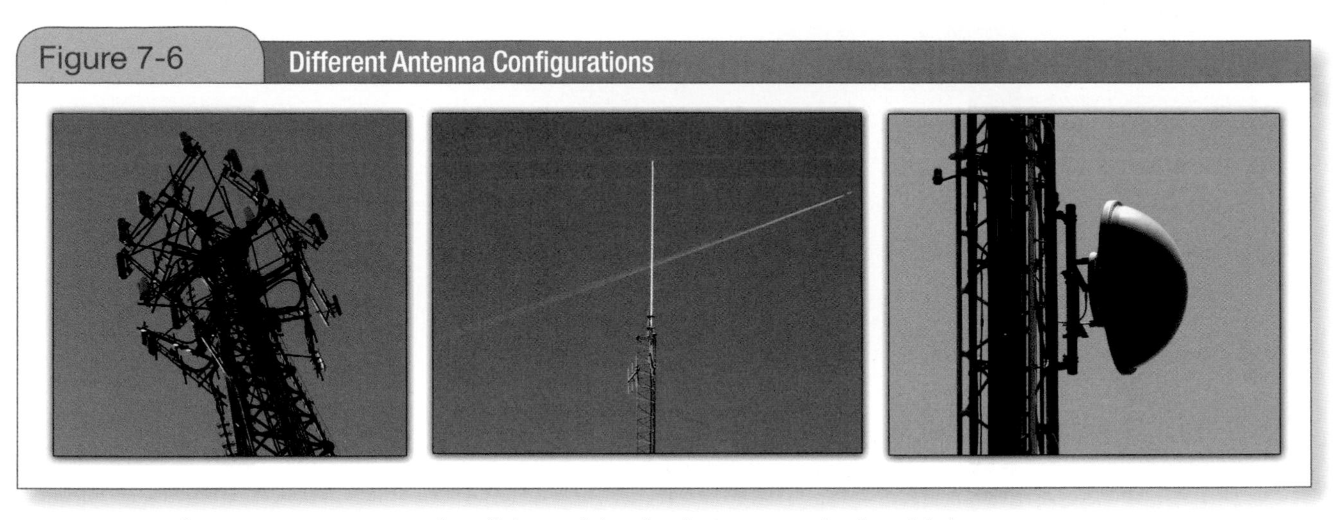

Figure 7-6. Article 810 covers antennas such as (left to right) multi-element, vertical rod, and dish antennas.

Article 810 consists of four parts. See the Table of Contents for a list of the Parts in this Article.

- Antenna systems for radio and television receiving equipment
- Amateur and citizen band radio transmitting and receiving equipment
- Transmitter safety
- Antennas and associated wiring and cabling – **See Figure 7-6**.

All of the following six parts of **Article 820** are dedicated to a single coverage: the cable distribution of radio frequency signals employed in community antenna television (CATV) systems. See the Table of Contents for a list of the Parts in this Article. **See Figure 7-7.**

Article 830 is divided into six parts to cover network-powered broadband communications systems that provide any combination of voice, audio, video, data, and interactive services through a network interface unit. **See Figure 7-8.**

Article 840 is divided into six parts covering premises-powered optical fiber-based communications systems that provide any combination of voice, audio, video, data, and interactive services through an optical network terminal (ONT).

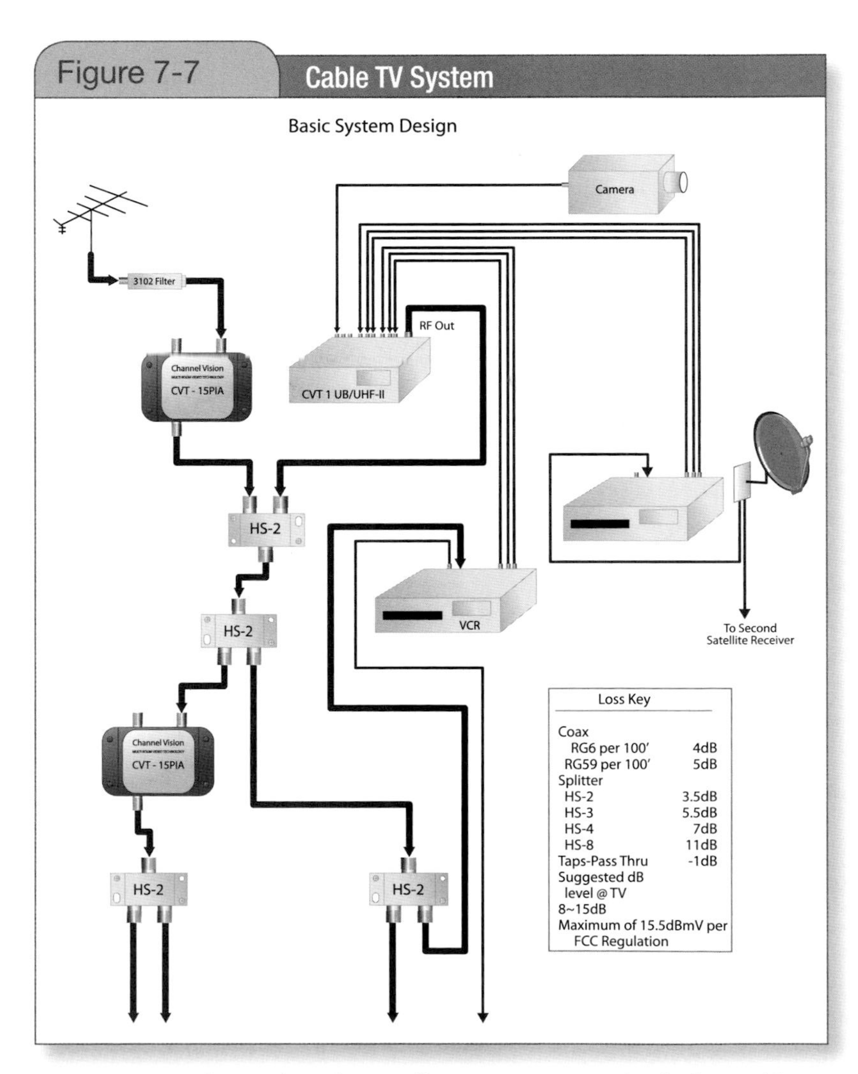

Figure 7-7. Article 820 lists the installation requirements for facility Cable TV systems.

Figure 7-8 Data Rack

Figure 7-8. Article 830 lists the installation requirements for facility data systems.

TABLES

As required in **Section 90.3**, Chapter 9 contains tables that apply as referenced throughout the *NEC*. **See Figure 7-9.** As such, they are extremely valuable tools for the *Code* user. A basic understanding of the types of tables in Chapter 9 and where they are referenced for use is necessary for quick reference to the correct table.

Table 1 Percent of Cross Section of Conduit and Tubing for Conductors

Table 1 is the benchmark for all permissible combinations of conductors in conduit and is commonly referred to as the conduit fill table. **Table 1** is short and to the point. It lists only three types of conductor installations and permitted percentage of raceway fill. **Table 1** is referenced in the *NEC* wherever raceway fill requirements exist. For example, when using rigid metal conduit, **Section 344.22** references the fill specified in **Chapter 9, Table 1. See Figure 7-10.**

Figure 7-9 *NEC* Chapter 9 Tables

***NEC* Chapter 9 Tables**	
Table 1	Percent of Cross Section of Conduit and Tubing for Conductors and Cables
Table 2	Radius of Conduit and Tubing Bends
Table 4	Dimensions and Percent Area of Conduit and Tubing (Areas of Conduit or Tubing for the Combinations of Wires Permitted in Table 1, Chapter 9)
Table 5	Dimensions of Insulated Conductors and Fixture Wires
Table 5A	Compact Copper and Aluminum Building Wire Nominal Dimensions* and Areas
Table 8	Conductor Properties
Table 9	Alternating-Current Resistance and Reactance for 600-Volt Cables, 3-Phase, 60 Hz, 75°C (167°F) - Three Single Conductors in Conduit
Table 10	Conductor Stranding
Table 11(A)	Class 2 and Class 3 Alternating-Current Power Source Limitations
Table 11(B)	Class 2 and Class 3 Direct-Current Power Source Limitations
Table 12(A)	PLFA Alternating-Current Power Source Limitations
Table 12(B)	PLFA Direct-Current Power Source Limitations

Figure 7-9. The tables in Chapter 9 are used to calculate conduit fill along with conductor properties in calculating conductor size due to voltage drop.

Figure 7-10 Notes to *NEC* Chapter 9, Table 1

Notes to *NEC* Chapter 9, Table 1

Table 1 Percent of Cross Section of Conduit and Tubing for Conductors and Cables

Number of Conductors and/or Cables	Cross-Sectional Area (%)
1	53%
2	31%
Over 2	40%

Table 1 Notes (Installation Requirements)

1. Fixture wire conduit fill calculation references to **Informative Annex C**.
2. **Table 1** requirements only apply to complete conduit installations and not applications of physical protection of exposed wiring.
3. Equipment grounding and bonding conductors are to be included in conduit fill calculations.
4. 60% conduit fill is permissible within 24 inches or less conduit installations between boxes, etc.
5. Actual outside dimensions of conductors not listed in Chapter 9 shall be used in conduit fill calculations.
6. **Tables 5 and 5A** shall be used for combinations of different size conductors in the conduit fill calculation and Table 4 for conduit and tubing dimensions.
7. The allowance to round-up to the next whole number in the case of conduit fill calculation with same size conductors
8. Requirements of **Table 8** for bare conductors
9. Allowances to count multiple-conductor cables as one conductor in the conduit fill calculations. If actual sizes are known, they are permitted to be used.

*Figure 7-10. Jamming can occur when pulling multiple conductors in a conduit. **Chapter 9, Table 1** lists the fill percentage of the conduit along with notes of installation requirements.*

Table 2 Radius of Conduit and Tubing Bends

Table 2 provides a uniform minimum requirement for bends in conduit (to prevent damage to raceways) and reduction of the internal area of conductors. **Table 2** is referenced in the raceway articles. For example, for cases using rigid metal conduit, **Section 344.24** references the bend requirements of **Chapter 9, Table 2**. Please note that in certain conduit installations, the job specifications may need a larger bend radius than the *NEC* requires. This occurs because of the type of conductor that may be installed, such as fiber optic cables or medium voltage cables. Larger-radius bends will reduce sidewall pressure in a cable pull, which could damage medium voltage cable. In the case of fiber optic cables, larger-radius conduit bends will prevent damage to the fiber and sustain the operation of light transmission to the fullest. **See Figure 7-11.**

Table 4 Dimensions and Percent Area of Conduit and Tubing (Areas of Conduit or Tubing for the Combinations of Wires Permitted in Chapter 9, Table 1)

Table 4 is extremely useful for the *Code* user when determining conduit fill. Information provided by this table includes total internal area and permissible fill area for several applications. **Table 4** is referenced in **Note 6** to **Table 1**, making **Table 4** applicable wherever **Table 1** is referenced in the *NEC*. **See Figure 7-12.**

Figure 7-11 Minimum Conduit Radius

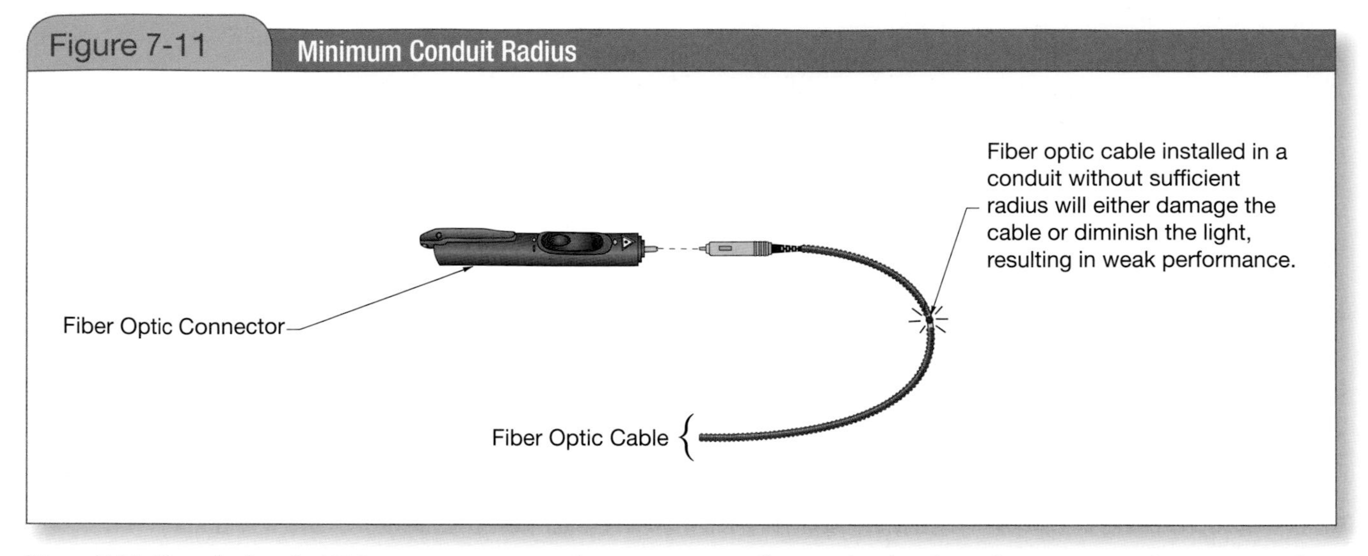

Figure 7-11. *Not only does the* NEC *require a minimum radius in raceway installations, often the job specifications increase the minimum radius.*

Table 5 Dimensions of Insulated Conductors and Fixture Wires

Table 5 provides the dimensions of insulated conductors and fixture wires needed to determine permissible conduit fill. **Table 5** is required to be used together with **Table 4** to determine permissible combinations of conduit fill. **Table 5** is referenced in **Note 6** to **Table 1**, making **Table 5** applicable wherever **Table 1** is referenced in the *NEC*.

Table 5A Compact Copper and Aluminum Building Wire Nominal Dimensions and Area

Table 5A provides the dimensions of compact copper and aluminum building wire needed to determine permissible conduit fill. **Table 5** is required to be used together with **Table 4** to determine permissible combinations of conduit fill. **Table 5A** is referenced in **Note 6** to **Table 1**, making **Table 5A** applicable wherever **Table 1** is referenced in the *NEC*.

Figure 7-12 Different Size Conductors in Same Raceway

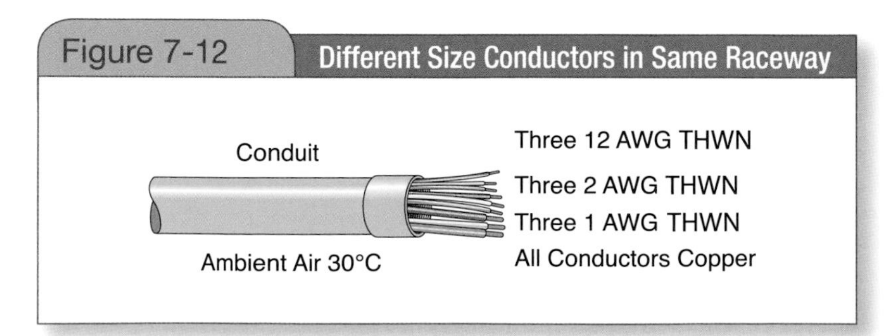

Figure 7-12. Table 4 *of Chapter 9 contains the information to calculate conduit size when filled with conductors of varying size.*

Table 8 Conductor Properties

Table 8 provides conductor properties for all conductor sizes from 18 AWG to 2000 kcmil. The information provided includes circular mil area for all AWG sizes, stranding, diameter, and DC resistance. **Table 8** is referenced in **Note 8** to **Table 1** for determining area for bare conductors, making **Table 8** applicable wherever **Table 1** is referenced in the *NEC*.

Table 9 Alternating-Current Resistance and Reactance for 600-Volt Cables, 3-Phase, 60 Hz, 75°C (167°F)—Three Single Conductors in Conduit

Table 9 provides resistance and impedance values necessary for determining proper conductor application where voltage drop or other calculations are required. In long run installations of conductors, voltage drop is directly proportional to the length from the source to the load. A good example is the long conductor runs to parking lot lighting. Typically, branch-circuit conductor runs to parking lot light fixtures can be several hundred feet. Using **Table 9** allows the user to calculate the increased conductor size in order to minimize the voltage drop to the lighting poles.

Table 10 Conductor Stranding

Table 10 declares the number of strands for Class B and Class C copper conductors and Class B aluminum conductors. There are five classifications of conductor stranding: Class AA, A, B, C, and D. Class AA is only a few strands, or practically a solid conductor, whereas Class D is the most finely stranded cable, such as a welding cable. The purpose of stranded cables in power applications is primarily for flexibility. The amount of strands in a cable typically determines the sizes of the strands; the more strands, the larger the outside circular dimension of the conductor. There are various reasons why the *NEC* declares the maximum amount of strands. One reason is related to the termination of a high-count stranded cable in a termination lug. If the strands are too small, they may not clamp down well in the termination.

Table 11(A) Class 2 and Class 3 Alternating-Current Power Source Limitations and Table 11(B) Class 2 and Class 3 Direct-Current Power Source Limitations

Table 11(A) and **Table 11(B)** provide the required power source limitations for listed Class 2 and Class 3 power sources.

Table 12(A) PLFA Alternating-Current Power Source Limitations and Table 12(B) PLFA Direct-Current Power Source Limitations

Table 12(A) and **Table 12(B)** provide the required power source limitations for listed PLFA power sources.

Figure 7-13 *NEC* Chapter 9 Informative Annexes

***NEC* Chapter 9 Informative Annexes**	
Informative Annex A	Product Safety Standards
Informative Annex B	Application Information for Ampacity Calculation
Informative Annex C	Conduit and Tubing Fill Tables for Conductors and Fixture Wires of the Same Size
Informative Annex D	Examples
Informative Annex E	Types of Construction
Informative Annex F	Availability and Reliability for Critical Operations Power Systems; and Development and Implementation of Functional Performance Tests (FPTs) for Critical Operations Power Systems
Informative Annex G	Supervisory Control and Data Acquisition (SCADA)
Informative Annex H	Administration and Enforcement
Informative Annex I	Recommended Tightening Torque Tables from UL Standard 486A-B
Informative Annex J	ADA Standards for Accessible Design

Figure 7-13. The Informative Annexes in Chapter 9 are a great source of vital information for the installer but are not part of the NEC *requirements.*

INFORMATIVE ANNEXES

As established in **Section 90.3**, Chapter 9 annexes are not part of the requirements of the *NEC* but are included for informational purposes only. A basic understanding of the types of annexes in Chapter 9 and the information they contain is necessary for the *Code* user to access this valuable information. **See Figure 7-13.**

Informative Annex A, Product Standards provides a list of UL, ANSI/ISA, and IEEE product safety standards used for the listing of products required to be listed in the *NEC*. These standards are very helpful for the installer for they not only provide vital information for safe installations but also sound installation practices to ensure maximum performance of the equipment.

Informative Annex B, Ampacity Calculation provides application information for many types of ampacity calculations, including those for conductors installed in electrical ducts.

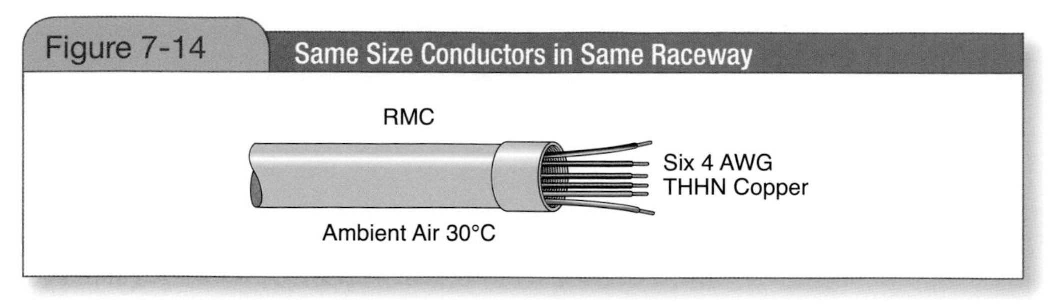

Figure 7-14. Informative Annex C of Chapter 9 contains the information to calculate conduit size when filled with conductors of the same size.

Informative Annex C, Tables is a useful aid to the user of the *Code* in determining conduit fill when all the conductors to be installed in a raceway are of the same size and type. **Informative Annex C** is informational only and is referenced in **Note 1** to **Table 1. See Figure 7-14.**

Informative Annex D, Examples is provided to aid the *Code* user when making calculations required by the *NEC*. The requirements for calculations are illustrated in the form of examples to aid those making similar calculations. **Informative Annex D** is a "must read" for all electrical installers.

For additional information, visit qr.njatcdb.org Item #1077

Informative Annex E, Construction Types is provided to aid the *Code* user when determining any of the five types of building construction. Several tables are included in **Informative Annex E,** which details fire resistance ratings, the maximum number of stories (floors) per type of construction, and cross-references.

The information in **Informative Annex F, Critical Operations Power Systems** is invaluable to the *Code* user when compliance with the commissioning requirements of **Article 708 Critical Operations Power Systems (COPS)** is necessary.

The information in **Informative Annex G, Supervisory Control and Data Acquisition (SCADA),** is useful when implementing a security control and data acquisition system that may be installed together with a critical operations power system described in **Article 708**.

Informative Annex H, Administration is provided as a model set of administration and enforcement requirements that could be adopted by a governmental body along with the electrical installation requirements of the *NEC*. Chapter 8 of this text discusses test preparation. Typically, the test is administrated by the local authority having jurisdiction (AHJ).

Informative Annex I, Tightening Torque Tables addresses torque. In the world of electrical installations, torque applications are all around you. Today, torque application includes everything from switches and breakers to connectors and busbar. Manufacturers typically have recommended torque requirements for proper installation. **Informative Annex I** lists recommended torque levels as defined in UL Standard 486A-B. Keep in mind, the manufacturer's torque recommendations supersede all other listed standards. **See Figure 7-15.**

Informative Annex J allows the *Code* user to properly consider electrical design constraints in the face of the 2010 ADA Standards for Accessible Design.

Figure 7-15 Torque Requirements Include Breakers

Figure 7-15. The manufacturer's torque requirements must supersede the torque recommendations listed in ***Informative Annex I*** *(UL Standards).*

Summary

Chapter 8, dedicated to communications systems, stands apart from the rest of the *NEC* in that the provisions of Chapters 1 through 7 apply to Chapter 8 only when a specific reference is made in a Chapter 8 article. Chapter 8 consists of the following five articles:

800 Communications Circuits
810 Radio and Television Equipment
820 Community Antenna Television and Radio Distribution Systems
830 Network-Powered Broadband Communications Systems
840 Premises-Powered Broadband Communications Systems

These five Chapter 8 articles cover specific methods, conductors, and equipment for communications systems. While other areas of the *NEC* may seem to be appropriately applied to a communications installation such as conduit or box fill, Chapter 8 is autonomous unless it specifically references any of the articles or sections in Chapters 1 through 7.

The twelve tables of Chapter 9 of the *NEC* apply only where referenced in the *Code*. These tables are necessary when applying many of the provisions of the *NEC*. Conduit fill calculations may include the use of several tables, including Table 1 through Table 8.

The *Code* Informative Annexes are intended only to aid the user of the *NEC*. These Informative Annexes offer useful information on product standards, conduit fill, types of construction, cross-reference tables, the application of **Article 708**, and administration and enforcement. Additionally, through the use of examples, **Informative Annex D** illustrates the calculations required in the *Code*.

Review Questions

1. **The tables in Chapter 9 of the *NEC* apply __?__.**
   a. as referenced in the *NEC*
   b. at all times
   c. only in Chapters 5, 6, and 7
   d. wherever they are useful
2. **Which table is referenced in raceway articles for conduit fill?**
   a. **Table 1**
   b. **Table 2**
   c. **Table 3**
   d. **Table 4**
3. **When the *NEC* requires wiring methods, materials, and equipment to be listed, Informative Annex __?__ provides additional information on the product standards.**
   a. A
   b. D
   c. F
   d. G
4. **The Informative Annexes in Chapter 9 of the *NEC* are __?__.**
   a. applicable as referenced
   b. mandatory requirements
   c. not mandatory requirements
   d. used only for special equipment
5. **Table 10 covers conductor __?__.**
   a. ampacity
   b. insulations
   c. properties
   d. stranding
6. **What other chapters apply generally to communication installations?**
   a. Chapters 1 through 4
   b. Chapters 1 through 7
   c. Chapters 5,6, and 7
   d. None of the other chapters apply.
7. **Chapter 8 is divided into how many parts?**
   a. 4
   b. 5
   c. 6
   d. 7
8. **The scope of Chapter 8 of the *NEC* is dedicated to circuits and equipment for all types of __?__ circuits.**
   a. communications
   b. data
   c. networking
   d. power
9. **What part of which article in Chapter 8 would address the installation of cables inside a building for cable TV?**
   a. **Part II**
   b. **Part III**
   c. **Part V**
   d. **Part VII**
10. **Article 830 is divided into how many parts?**
    a. 2
    b. 3
    c. 5
    d. 6
11. **What part of Article 830 covers grounding methods for network-powered broadband communications systems?**
    a. **Part II**
    b. **Part III**
    c. **Part IV**
    d. **Part V**
12. **Which article in Chapter 8 covers requirements for amateur radio antenna systems?**
    a. **800**
    b. **810**
    c. **820**
    d. **830**

Test Preparation

Introduction

The key to success in any type of exam is to be fully prepared. However, a *Code* exam can be quite different from other exams that applicants may have taken in the past. The ability to quickly and accurately find information in the *NEC* using the *Codeology* method as well as sound time management are the keys to success when taking an *NEC* proficiency exam. In addition, users must be able to recognize the various types of exams, questions, and testing methods before approaching test day with any level of self-confidence.

Objectives

» Recognize the importance of the *Codeology* method when taking *NEC* exams.

» Identify the different types of exams.

» Describe the different types of exam questions.

» Recognize the importance of proper time management when taking an exam.

» Recognize the importance of developing a game plan before taking an exam.

Chapter 8

Table of Contents

For additional information, visit qr.njatcdb.org Item #2641

EXAMS BASED UPON THE *NEC* FOR ELECTRICAL CONTRACTOR, JOURNEYMAN OR MASTER ELECTRICIAN, AND INSPECTOR

In many states, electrical contractors, electrical inspectors, and Electrical Workers are required to be licensed. Preparing for an *NEC* exam to become an electrical contractor, inspector, or worker requires many hours of study.

The three prerequisites for those about to take an *NEC* exam are preparation, preparation, and more preparation.

A method to quickly and accurately find the needed information is necessary for success when taking an *NEC* exam. The *Codeology* method is the most valuable tool for use on test day. Time management is the second most valuable tool.

Article-specific courses are necessary to obtain proficiency in areas such as grounding and bonding, along with mastering calculations for services and feeder sizing, box and conduit fill, conductor ampacity corrections, and motor installations. All *NEC* exams are divided into groups of questions, the majority of which can be answered without time-consuming calculations. For example, in an exam of 80 questions, perhaps 50 or more will not require time-consuming calculations. These 50 questions are those which can be quickly and accurately answered using the *Codeology* method. These questions are often closed book. The remaining questions are typically open book and can be answered by doing the required calculations after locating the correct requirements in the *NEC*. Using the *Codeology* method for the calculation questions will be vital.

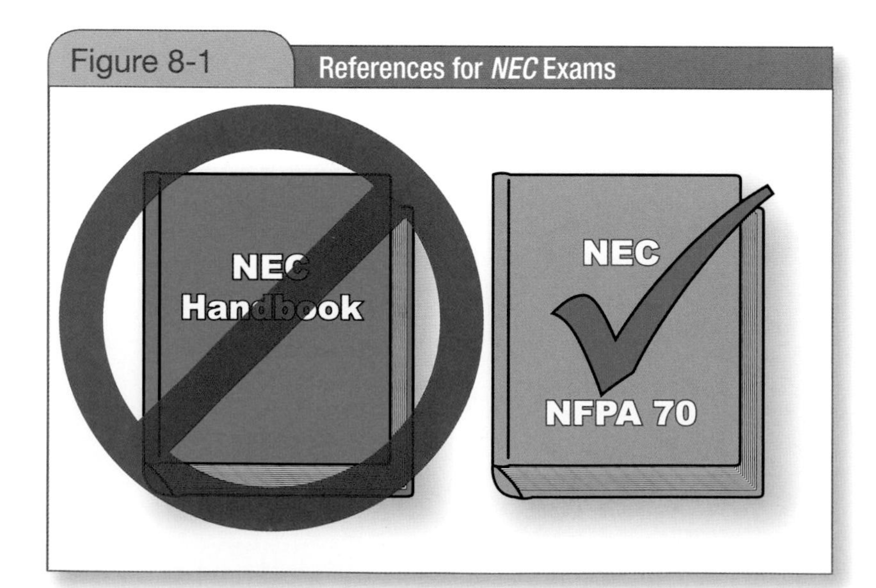

Figure 8-1. Most NEC *exams do not allow the* NEC *Handbook as a reference in the test. Often, notes written in the margins of the* NEC *are also not allowed.*

TYPES OF EXAMS

Preparing for the exam includes contacting the testing organization and requesting information such as dates, times, submissions required prior to the day of testing, fees, and what materials are allowed to bring into the testing room. Of course, the *NEC* will be allowed for the test, but many test situations do not allow the ***NEC* Handbook**. **See Figure 8-1.** The ***NEC* Handbook** contains the entire Electrical *Code* (***NFPA 70***) with additional informative information to give the reader a clear explanation of the *Code*. In some cases, a *Code* book that contains highlights and written notes is not allowed. Most likely, an electrical calculation calculator will not be allowed. Be sure to confirm what is and is not allowed in respect to electrical theory textbooks as well. It would be unfortunate to arrive at the testing center only to be unable take the test because a personal copy of the *Code* book contained tabs.

Before the test, make sure to verify whether or not the following items are allowed:

- Is the *Code* book, Handbook, or either allowed?
  - May it be highlighted or tabbed?
  - Are any notes allowed in the reference material?
- Other References:
  - What are the allowed references other than the *NEC*?
  - Formula references and books?
  - Local Codes or Ordinances?
  - May any of these be highlighted or tabbed?

- Calculator
  - Programmable or not?
  - Scientific or not?
  - Electrical calculator or not?
- Tabs
  - Must they be published or can the applicant make his/her own?
  - May they be handwritten or typed?
  - Must they be permanently attached?

Written Exams

Written exams are slowly becoming a thing of the past due to the advantages of taking an online exam proctored at a local licensing facility. Nevertheless, the written test, which utilizes a scanned answer form, is still used in some areas. A written exam usually consists of an exam sheet containing multiple-choice questions and an answer scan sheet. The submitted answers are not collected from the exam sheet, but rather from the answer scan sheet.

Most applicants will use the exam sheet to perform their calculations and even mark the answer for reference, but the final answers must be marked on the answer scan sheet by filling in one of the four possible answer circles for each question. Once the testing session is over, the test administrator will deliver the answer scan sheets to the testing organization for scanning and determining grades. **See Figure 8-2.**

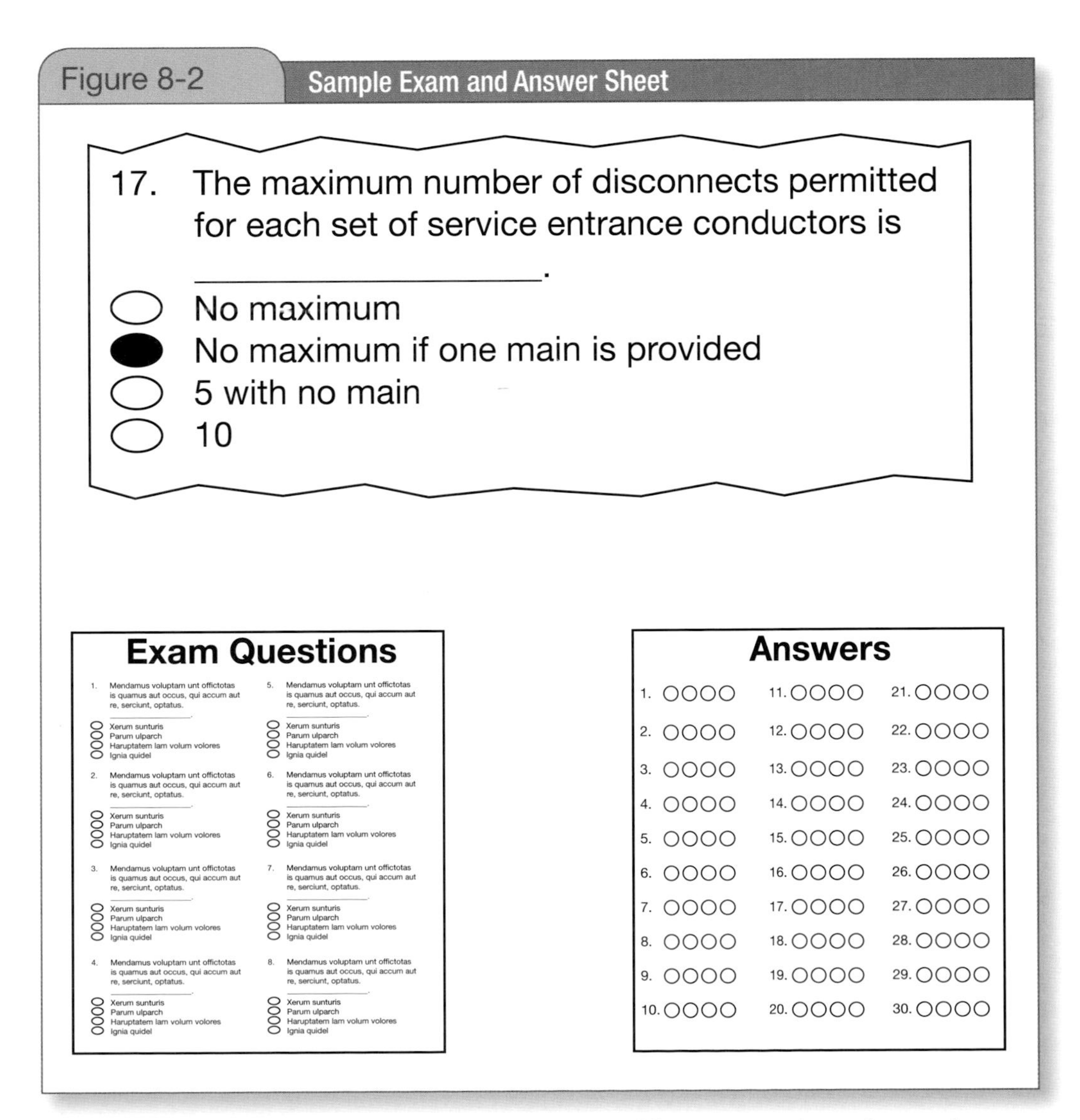
Figure 8-2
Sample Exam and Answer Sheet
17. The maximum number of disconnects permitted for each set of service entrance conductors is ________.
No maximum
No maximum if one main is provided
5 with no main
10
Exam Questions
Answers

Figure 8-2. Typically, a written exam requires the applicant to fill in answer circles on the scan sheet.

When taking a written exam, it is vital that an applicant place each answer in its proper location. For example, if an applicant accidentally skips Question No. 2 on the exam sheet and fills in the answer for Question No. 3 in the space intended for Question No. 2, the exam would be completed with answers one question off on the answer scan sheet. This would be a real waste of the applicant's time.

Computer Exams

Computer-based exams are easily navigated for the computer user who possesses basic computer skills. On the other hand, applicants without basic computer skills should seek assistance and practice using a PC before test day. Most providers of *NEC* exams will have sample questions available on their websites. All applicants, with or without basic computer skills, should take advantage of the opportunity to become familiar with the computer testing format. Additionally, in most cases, the exam administrator will offer a short tutorial on the exam format before the exam. One of the advantages of computer exams is receiving performance grades shortly after the exam session.

Figure 8-3 Computer Exam

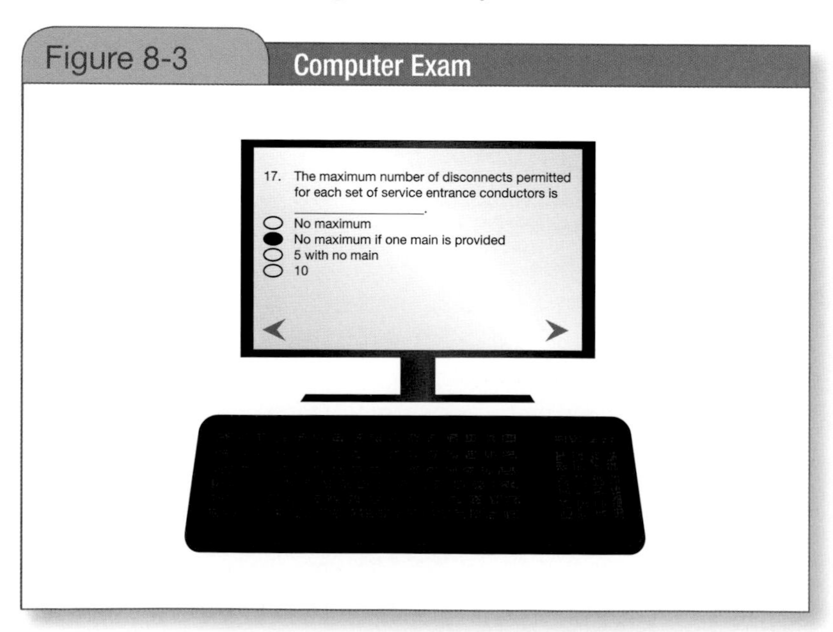
17. The maximum number of disconnects permitted for each set of service entrance conductors is ________.
No maximum
No maximum if one main is provided
5 with no main
10

Figure 8-3. The computer exam is the most popular method of taking the NEC *exam today.*

Applicants will still need to bring paper to perform the calculations before entering the answer on the computer screen. **See Figure 8-3.**

READING EACH QUESTION

Applicants should read each multiple choice question and all of the answers carefully before applying the *Codeology* method. It is essential that the question and all of the answers be read because a keyword or clue may be in one of the possible answers. If the answer is not apparent, then consider highlighting the question to be solved later. By reading each question and all of the answers, the applicant will subconsciously retain those questions. As applicants move through the exam, other questions and/or answers may jog the memory to solve a skipped question.

TYPES OF QUESTIONS

Code questions can be broken down into two basic categories: those which can be solved quickly and may involve a simple calculation, and those which require time-consuming calculations.

Understanding this concept and applying the proper time management techniques throughout the exam will increase the chances of success.

General Knowledge Questions

General knowledge questions are included in all *NEC* exams. These questions are not designed to determine *NEC* proficiency, but to determine general knowledge of the electrical trade and electrical theory. General knowledge questions are often part of the closed book portion of the exam.

General questions often include the following material:

- Ohm's Law
- Power formula
- AC formulas
- Inductance and capacitance
- Conversions (for example, horsepower to watts)

- Voltage drop
- Local codes

Multiple Choice Questions

Typically, four answers follow each question in an *NEC* exam. When the answer is known, choose the correct option and move on to the next question. All questions must be answered by the end of the exam unless there is no penalty for unanswered questions. For multiple-choice questions which are difficult to answer, try to eliminate one or more of the options. When more than one answer seems to be correct and "all of the above" is an option, it is probably the correct answer. If only two of the four possibilities can be eliminated and the applicant is forced to guess which of the remaining two is correct, the odds of getting a correct answer are still significantly increased. **See Figure 8-4.**

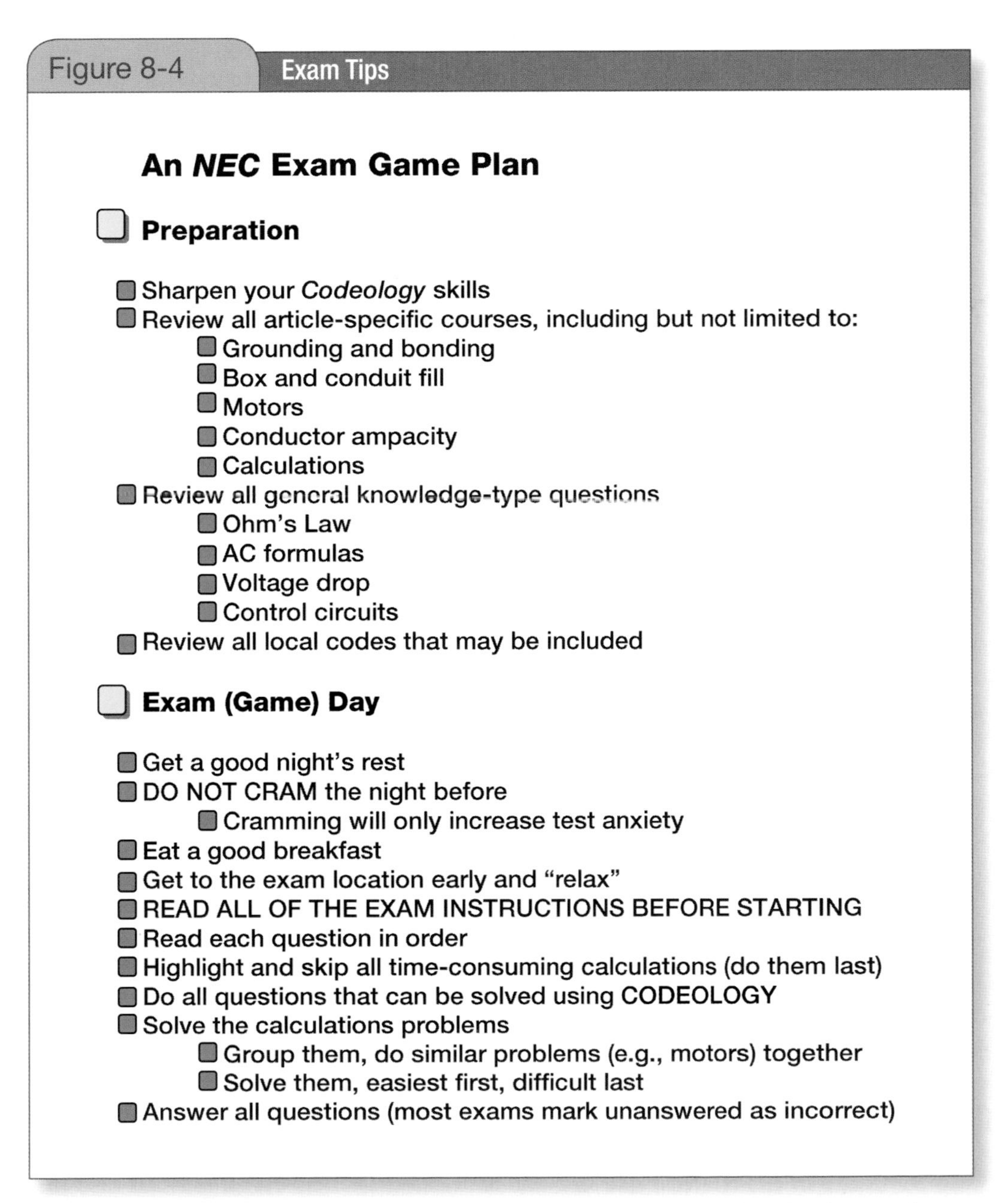

Figure 8-4
Exam Tips
An NEC Exam Game Plan
Preparation
Sharpen your Codeology skills
Review all article-specific courses, including but not limited to:
Grounding and bonding
Box and conduit fill
Motors
Conductor ampacity
Calculations
Review all general knowledge-type questions
Ohm's Law
AC formulas
Voltage drop
Control circuits
Review all local codes that may be included
Exam (Game) Day
Get a good night's rest
DO NOT CRAM the night before
Cramming will only increase test anxiety
Eat a good breakfast
Get to the exam location early and "relax"
READ ALL OF THE EXAM INSTRUCTIONS BEFORE STARTING
Read each question in order
Highlight and skip all time-consuming calculations (do them last)
Do all questions that can be solved using CODEOLOGY
Solve the calculations problems
Group them, do similar problems (e.g., motors) together
Solve them, easiest first, difficult last
Answer all questions (most exams mark unanswered as incorrect)

Figure 8-4. To be successful on the NEC exam, the applicant should review the related materials over several weeks instead of cramming a few days prior.

TIME MANAGEMENT

When taking an *NEC* exam, a given number of questions must be answered within a specified amount of time. For example, an 80-question exam could be given in a four-hour period. Proper time management is essential when taking any exam. Additionally, most exams assign the same weight, or value, to each question. For example, in an 80-question, 4-hour exam, each question would be worth 1.25 points with only three minutes to solve each question. Time-consuming calculations may take ten minutes or more while others can be solved in one minute or less using the *Codeology* method.

Thus, when taking an *NEC* exam, time-consuming calculations should always be skipped and solved last, regardless of how comfortable the applicant may feel in solving the problem. Interviews of applicants after taking an *NEC* exam typically reveal the following two scenarios:

Scenario 1

Marty is taking a four-hour, 80-question electrical contractor exam. He solves each question in order. The front of the test is loaded with calculation-type questions. Marty is hung up on a few questions and his calculations do not match any of the answers. He checks his watch and realizes that 2 hours have passed, and he is only on Question No. 18, leaving only two hours to complete the remaining 62 questions. Panic sets in and Marty starts getting nervous. His self-confidence has dissipated, and he begins to believe that he will fail the exam. Marty's ability to calmly read and solve each question is severely compromised. He did not use his time wisely.

Scenario 2

Tamara is taking a four-hour, 80-question electrical contractor exam. She reads each question in order, highlighting the time-consuming, calculation-type questions for completion later and uses *Codeology* to quickly solve the other questions. Tamara checks her watch and realizes that 2 hours have passed, and she has completed 55 questions, leaving two more hours to complete the remaining 25 questions. She is pleased with her progress. Tamara's self-confidence has grown, and she feels sure she will pass the exam. Tamara's ability to calmly read and solve each question has increased because of her renewed self-confidence. Unlike Marty, she has used her time wisely.

Summary

Preparing for and taking any *Code* exam, whether for an entry-level position or to become an electrical contractor, can be an unnerving experience for many Electrical Workers. Without a game plan, time management skills, and a method to quickly solve basic questions, the chances of passing the exam are not high for the ill-prepared applicant.

Proficiency in the *NEC* requires more than *Codeology* and time-management skills. For most, it also requires taking article- or topic-specific *Code* courses to become competent in calculations, grounding, and other areas. However, on test day, the two most important tools for the applicant are:

- A method to quickly find needed information (*Codeology*)
- Time management

Applicants must familiarize themselves with the type of upcoming test, either written or computer-based. Applicants must also understand the type of questions to be included

Summary

on an exam, the value of each question, the number of questions, the time allowed, and whether unanswered questions will be marked as incorrect. Time-consuming calculations should be left for completion at the end of the exam. Whether the exam is written or on a computer, the applicant must contact the testing organization and request information such as dates, times, submissions required prior to the day of testing, fees, and what reference materials can be allowed in the testing room.

Proper preparation for an exam will result in a high level of self-confidence, allowing the applicant to be more relaxed. Skipping time-consuming problems early in the exam and using *Codeology* to solve all the others results in increased self-confidence. It will also allow ample time near the end of the exam to solve calculations calmly and carefully.

Review Questions

1. *Code* users taking a competency exam on the *NEC* must take which of the following steps to be completely prepared?

a. Be capable of quickly and accurately finding information
b. Practice sound time management
c. Prepare and study for all types of *Code* questions
d. All of the above

2. Using proper time management methods, the *Codeology* user will skip all the time-consuming __?__ problems and solve them after all other questions have been answered.

a. calculation
b. Chapter 5
c. grounding
d. multiple-choice

3. General knowledge-type questions on *NEC* competency exams may include questions involving __?__.

a. control circuits
b. Ohm's Law and AC formulas
c. voltage drop
d. all of the above

4. Lack of proper preparation such as time management skills, the ability to quickly and accurately find information, and article-specific preparation will result in reduced __?__ and ultimately, failure on an exam.

a. ability to guess
b. good luck
c. self-confidence
d. speed reading

5. Which of the following is/are not likely to be permitted in the testing room?

a. *Code* book with notes written in the margins
b. Electrical calculation calculators
c. *NEC* Handbook
d. All of the above

Making of a Code: The *National Electrical Code* Process

Introduction

The *National Electrical Code* (*NEC*) is revised by the National Fire Protection Association (NFPA) every three years and made available for use by designers, code enforcers, contractors, installers, and other *Code* users. Three years prior to a new release of an NFPA document, NFPA Technical Committee (TC) members and staff collaborate extensively with task groups to consider changes presented to the *Code*-making panels (CMPs) by *Code* users and by the public and to address other issues identified in the previous change cycle. Special task groups work on technical issues and usability issues for consideration by the CMPs. The *NEC* is a dynamic document, a true work in progress that is shaped, molded, and improved by all members of the public who take part in the process.

Appendix A

Table of Contents

THE *CODE*-MAKING PROCESS

This text, *Applied Codeology*, is an introduction to understanding the *NFPA 70: National Electrical Code* (*NEC*). The NFPA was founded in 1896 in Boston by a small group of men who saw the need to set standards for protecting the public against fire hazards. The name has been altered throughout the years, but the initial goal remains the same—protection of the public. Initially, the organization developed recommended practices for fire suppression using sprinklered water and addressed the design and installation of early electrical distribution systems, but over the past 100 years, the NFPA has become the code-making organization for the fire protection codes that most U.S. municipalities adopt.

Q. *How are the* Codes *derived and designed?*

A. *The NFPA* Code*-making process is not achieved exclusively within the NFPA organization—quite to the contrary. NFPA codes are completely developed by volunteer users of the document. All the various NFPA codes have an associated* Code*-making panel (CMP) which is closely affiliated with that particular code area. Members assigned to each CMP work collaboratively to research, define, and submit the* Code *language for public approval. Members of the electrical industry in the United States completely drive the language that makes up the* NEC.

Structure of the NFPA

The NFPA derives minimum codes and standards for fire prevention and other life-safety codes and standards for the United States. The NFPA Board of Directors is the top governing body over all of the 3001 codes and standards that the NFPA maintains. The Board of Directors appoints a Standard Council of 13 members that oversees and administers the rules and regulations. The Standard Council appoints all CMPs, TCs, Correlating Committees (CCs), and Motion Committees. In addition, the Standard Council derives new CMPs and TCs if new technologies or situations warrant them. Both the NFPA and *NEC* contain a basic hierarchy of governing bodies. **See Figure A-1.**

Among the 3001+ codes and standards are several electrical-related chapters, such as *NFPA 70: National Electrical Code*, *NFPA 70E: Standard for Electrical Safety in the Workplace*, and *NFPA 72: National Fire Alarm and Signaling Code*. There are several NFPA codes that the Electrical Worker may be required to review during the layout or installation stages of a project. **See Figure A-2.** Keep in mind that all 300+ codes and committees are revised every three years, but are not on the same three-year schedule. For instance, the *NFPA 70* (*NEC*) schedule is 2014, 2017, and so on, while the *NFPA 72* schedule is 2013, 2016, and so on.

NEC Process

The *NEC* (*NFPA 70*) is developed through a *consensus* standards development process approved by the American National Standards Institute (ANSI). This process includes input from all interested parties (public) throughout the proposal, comment, and amending-motions stages. TCs, also known as CMPs, are formed to achieve *consensus* on all proposed changes or revisions to the *NEC*. These volunteer committees, with members representing various classifications, are balanced to ensure all viewpoints of all interest groups have a voice in the deliberations of issues brought before the *NEC*.

As defined in the NFPA *Manual of Style for NFPA Technical Committee Documents,* consensus, as applied to NFPA documents, means that a substantial agreement to a proposed change of language in a *Code* or standard has been reached by the affected interest categories. Substantial agreement means much more

Figure A-1 Hierarchy of Governing Bodies

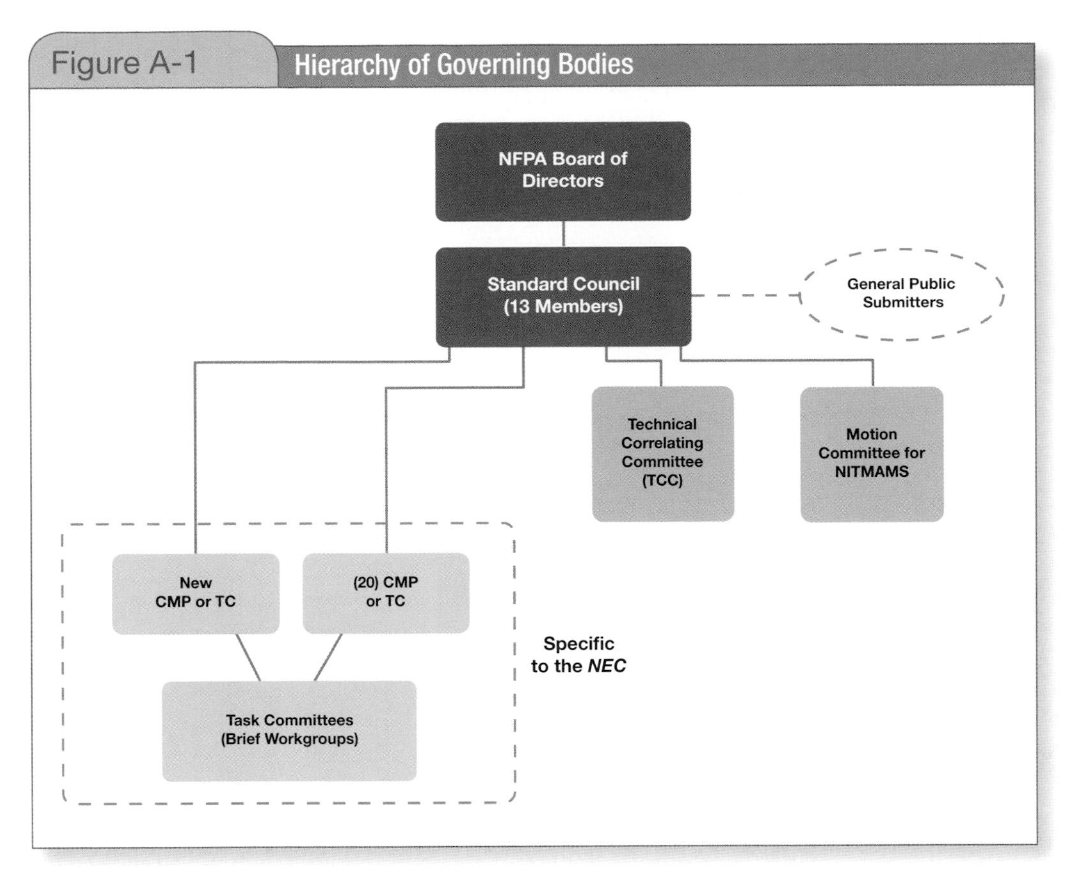

Figure A-1. The NFPA Representative Structure is the same throughout all the Standards and Codes.

than a simple majority; two-thirds of the TC members must agree, but the vote to adopt a proposal does not necessarily need to be unanimous.

It is important for all users of the *NEC* to understand the document revision process. This process allows for public participation in the development of the *NEC*. The *NEC* is revised every three years. Each revision cycle has an established schedule published in the back of the *Code* book that includes final dates for all the major stages of the revision process. The process has four basic steps:

1. Input Stage
2. Comment Stage
3. Association Technical Meeting
4. Council Appeals and Issuance of Standard:

Throughout the revision cycle, the process produces the *First Draft Report* (FDR), previously known as the Report on Proposals (ROP); a *Second Draft Report* (SDR), previously known as Report on Comments (ROC); and Certified Amended Motions for the Association Technical Meeting (NFPA National Meeting). The older method using ROP and ROC was used with the 2014 edition of the *NEC*. The newer method is used starting with the 2017 *NEC*. The TCs work diligently to provide the *NEC* user with *Code* text that is practical, easy to read and understand, and enforceable. However, as *Code* changes are applied, users may disagree about the implementation of some of the revised requirements. The intent of the changes can be found in the substantiations, CMP statements, and CMP member comments at both the FDR and SDR stages. All persons submitting a proposal are notified of the committee's action and have access to the *First Draft Report* and *Second Draft Report* on the NFPA's website. These documents are extremely valuable for the user of the next edition of the

Figure A-2	Partial (Electrical) Listing of NFPA Codes and Standards
NFPA 1	Fire Code
NFPA 3	Recommended Practice on Commissioning and Integrated Testing of Fire Protection and Life Safety Systems
NFPA 20	Standard for the Installation of Stationary Pumps for Fire Protection
NFPA 30A	Code for Motor Fuel Dispensing Facilities and Repair Garages
NFPA 70	National Electrical Code®
NFPA 70A	National Electrical Code® Requirements for One- and Two-Family Dwellings
NFPA 70B	Recommended Practice for Electrical Equipment Maintenance
NFPA 70E	Standard for Electrical Safety in the Workplace®
NFPA 72	National Fire Alarm and Signaling Code
NFPA 73	Standard for Electrical Inspections of Existing Dwellings
NFPA 75	Standard for the Fire Protection of Information Technology Equipment
NFPA 76	Standard for the Fire Protection of Telecommunications Facilities
NFPA 79	Electrical Standard for Industrial Machinery
NFPA 92	Standard for Smoke Control Systems
NFPA 96	Standard for Ventilation Control and Fire Protection of Commercial Cooking Operations
NFPA 99	Health Care Facilities Code
NFPA 101	Life Safety Code®
NFPA 110	Standard for Emergency and Standby Power Systems
NFPA 111	Standard on Stored Electrical Energy Emergency and Standby Power Systems
NFPA 170	Standard for Fire Safety and Emergency Symbols
NFPA 220	Standard on Types of Building Construction
NFPA 262	Standard Method of Test for Flame Travel and Smoke of Wires and Cables for Use in Air-Handling Spaces
NFPA 450	Guide for Emergency Medical Services and Systems
NFPA 496	Standard for Purged and Pressurized Enclosures for Electrical Equipment
NFPA 497	Recommended Practice for the Classification of Flammable Liquids, Gases, or Vapors and of Hazardous (Classified) Locations for Electrical Installations in Chemical Process Areas
NFPA 499	Recommended Practice for the Classification of Combustible Dusts and of Hazardous (Classified) Locations for Electrical Installations in Chemical Process Areas
NFPA 654	Standard for the Prevention of Fire and Dust Explosions from the Manufacturing, Processing, and Handling of Combustible Particulate Solids
NFPA 730	Guide for Premises Security
NFPA 731	Standard for the Installation of Electronic Premises Security Systems
NFPA 780	Standard for the Installation of Lightning Protection Systems
NFPA 791	Recommended Practice and Procedures for Unlabeled Electrical Equipment Evaluation
NFPA 850	Recommended Practice for Fire Protection for Electric Generating Plants and High Voltage Direct Current Converter Stations
NFPA 853	Standard for the Installation of Stationary Fuel Cell Power Systems
NFPA 5000	Building Construction and Safety Code®

Figure A-2. Although NFPA 70 is most familiar to the electrician, there are many other NFPA codes that apply to electrical installations.

For additional information, visit qr.njatcdb.org Item #1060

NEC. In the back of each edition of the *NEC*, there are instructions for participating in the *Code* change process. Anyone can submit a proposed change to the next edition of the *NEC*. Most proposals will be submitted online at the National Fire Protection Association (NFPA) website, www.nfpa.org.

SEQUENCE OF EVENTS

Starting with the 2017 *NEC* cycle, the NFPA moved to a new format for updating the *National Electrical Code*. The following explains the consensus standards development process. Actual deadlines for each of these steps are listed on the NFPA website and in the back of the *NEC*. **See Figure A-3.**

Public Input

One of the first steps in the change process is the public input stage. This process begins when a person wishing to submit a code change accesses the Standards Development Site on the NFPA web page, where there is a tutorial and clear instructions for submitting changes.

A public notice requesting interested parties to submit specific written proposals is published in the NFPA News, the U.S. Federal Register, the American National Standards Institute's Standards Action, the NFPA's website, and other publications. In this stage, all proposals are submitted electronically on the NFPA's website. Proposed changes to the *Code* are submitted to the NFPA, which assigns them to a TC or a CC for consideration in the next edition. As shown in the *NEC*, a CMP is made up of volunteers to review code-change proposals.

The public inputs are created by the interested party and submitted electronically under the following categories:

- New additions to the *Code*
- Revision of an existing section
- Creation of a global revision to add, modify, or delete a word or phrase throughout the entire document.

Once a topic for changing the *Code* has been selected, the table of contents in the *NEC* appears to help navigate to the chapter, part, article, and section that addresses the change. All changes are shown in proper editorial format to ensure clarity. The submitter can save, change, or delete a proposal at any time before the public input submission deadline date.

For additional information, visit qr.njatcdb.org Item #1061

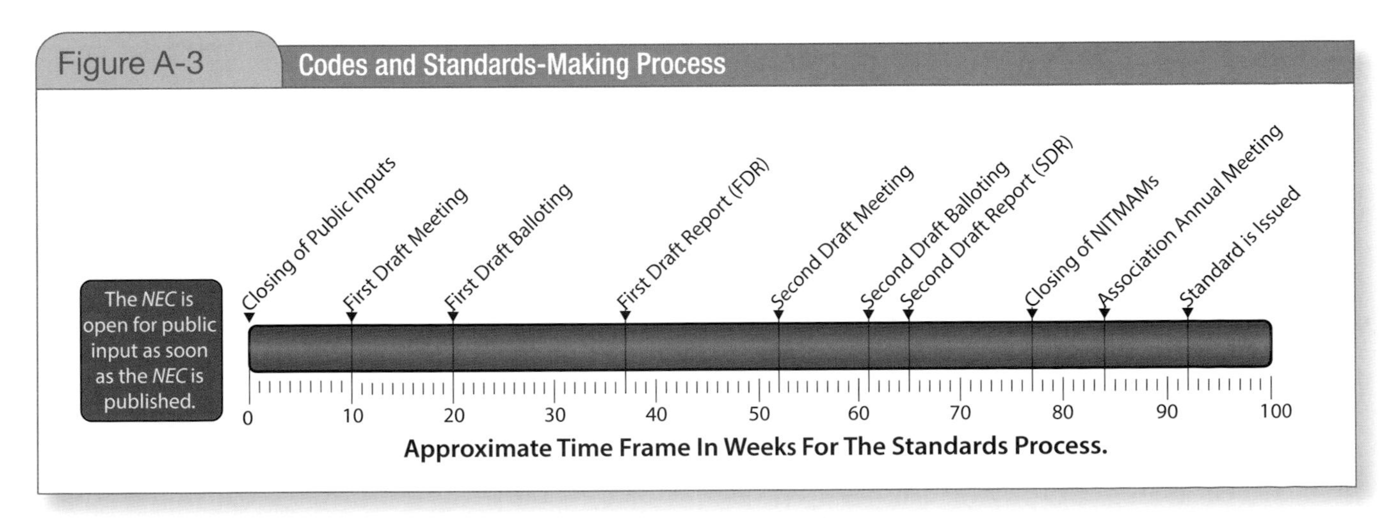

Figure A-3. There are many steps to Code-making during the three-year Code cycle.

Input Stage – First Draft Report (FDR)

Once the Public Input date has passed, the proposals are forwarded to the CMP to be addressed at the Public Input meeting. In this public meeting, the CMP reviews and votes on the proposals and develops the first draft of the *Code*. The CMPs meet at the location and dates stated in the back of the *NEC* to act on all public input. These meetings are open to all interested persons who want to observe the panel proceedings. Actions at the meetings require only a simple majority vote.

After completion of the public input meeting and balloting, the first draft report is issued, and the submitter is notified of the committee's action. The committee provides an action and response to each public input.

After the CMP meetings for the first draft have closed, the ballots are sent to all panel members. A two-thirds majority on the ballot is required for the panel action to be upheld. All ballots are filed electronically. The first draft report is available to the public, submitters, and CMPs.

Public Comment Stage and the Second Draft Report

Next is the public comment stage. Once the first draft report becomes available, there is a public comment period during which anyone may submit a public comment on the first draft. Any objections or further related changes to the content of the first draft must be submitted at the comment stage; however, no new material is allowed to be considered.

After the public comment closing date, the committee holds their second draft meeting. All the public comments are considered, and the committee provides an action and response to each public comment. The committee will use the public comments in order to help develop second revisions to the standard, which results in a complete and fully integrated draft known as the *second draft report.* Like the first draft, the second draft has initial agreement by the committee based on a simple majority vote during the meeting to establish a consensus. The final position of the committee is established by ballot.

Next, the committee votes by ballot on the second draft. The second revisions developed at the second draft meeting are balloted. This means that the text the committee wants revised in the standard is on the ballot for approval by the committee. It is important to point out that in the second draft meetings only a majority vote is required to move to the final ballot. Second revisions must be approved by two-thirds of the committee to appear in the second draft. Any second revisions that do not pass the ballot appear in the second draft report as committee comments.

The second draft report is posted on the NFPA website. The second draft report serves as documentation of the comment stage and is published for public review.

NFPA Annual Association Meeting and Amending Motions

The next step in the process is the Annual Association Meeting. Following the completion of the input and comment stages, there is further opportunity for debate and discussion of issues through the Association Technical Meeting (Technical Meeting), which takes place at the NFPA Conference & Expo. The full *NEC* TC report is presented to the NFPA membership for approval at the NFPA Annual Association Meeting. Because the *NEC* revision is made available to the public for revision years 2017, 2020, and so on, the annual meetings that include the *NEC* amending motions and approvals are held during the mid years of 2016, 2019, and so on.

Motions may be made to amend or reverse the actions taken by the CMPs at the annual meeting. Before making a motion at the Annual Association Technical Meeting, the intended maker of the motion must file a notice of intent to make a motion (NITMAM) 30 days in advance. The cycle schedule will list the deadline date to submit a NITMAM before the annual meeting. The NITMAM will be received and approved by the Motions Committee and listed on the approved motion listing as Certified Amending Motions (CAMs) for the annual meeting. At the annual meeting, the submitter (or representative) of the NITMAM must notify the NFPA one hour prior to the meeting start time that they will be making the motion. Debate and voting by NFPA members will be allowed at the annual meeting on all CAMs and will carry with a majority vote.

Appeals to the Standards Council

Appeals to the Standards Council may be made up to 20 days after the annual meeting, after which the Standards Council adjudicates any appeals, accepts the new *NEC*, and issues the revised *Code*.

NEC COMMITTEES

In addition to the Standards Council, the NFPA uses specific committees in the development of the *NEC*. These committees collaborate in a seamless structure to ensure that the newly developed code proposals are best suited throughout all the NFPA codes. The committees also perform technical research to better understand the future needs of public safety within the electrical industry. Some of these committees are maintained as part of the standard *Code*-making process, while other committees may be briefly convened to perform a particular task and then dissolved shortly afterward.

Correlating Committee

The Correlating Committee (CC) reviews all of the CMP results to ensure that there are no conflicting actions. Their duties are extensive because they are required to review multiple CMPs' second draft reports. Typically, the chairmen of the TCs will advise the CC to ensure that possible conflicting actions are resolved. **See Figure A-4.**

Code-Making Panels

Presently, the *NEC* CMPs consist of nineteen panels. As new technologies develop, the Standard Council will recognize the need to derive new CMPs. **See Figure A-5.**

The IBEW and *NECA* are represented on many of the NFPA CMPs. All the *NEC* panels and members are listed in the front of the *NEC*. Likewise, for other NFPA Codes and Standards (for example, *NFPA 72*), the CMP members are listed in the front of the publication. Take time to review the CMP listing to get a sense of who is occupying the CMPs and acting as the driving force behind the *Code*-making process. **See Figure A-6.**

COMMITTEE MEMBERSHIP

Members of the NFPA committees are made up of subject-matter experts who are employed or associated with the subject of the committee. As discussed earlier, the committees are made up of volunteer users of the *Code* such as research/testing labs, enforcing authorities, insurance agencies, consumers, manufacturers, utilities, and special experts.

To become a member of the NFPA, the applicant is required to complete a form that requests a variety of information such as qualifications, relationship to other members, availability to participate actively, funding source for the applicant's participation, background of the applicant's employer, and other

Figure A-4 **2017 *NEC* Correlating Committee**

NATIONAL ELECTRICAL CODE COMMITTEE

These lists represent the membership at the time the Committee was balloted on the final text of this edition. Since that time, changes in the membership may have occurred. A key to classifications is found at the back of this document.

Correlating Committee on National Electrical Code®

Michael J. Johnston, *Chair*
National Electrical Contractors Association, MD [IM]

Mark W. Earley, *Secretary (Nonvoting)*
National Fire Protection Association, MA

Kimberly L. Shea, *Recording Secretary (Nonvoting)*
National Fire Protection Association, MA

James E. Brunssen, Telcordia Technologies (Ericsson), NJ [UT]
Rep. Alliance for Telecommunications Industry Solutions

Kevin L. Dressman, U.S. Department of Energy, MD [U]

Palmer L. Hickman, Electrical Training Alliance, MD [L]
Rep. International Brotherhood of Electrical Workers

David L. Hittinger, Independent Electrical Contractors of Greater Cincinnati, OH [IM]
Rep. Independent Electrical Contractors, Inc.

Richard A. Holub, The DuPont Company, Inc., DE [U]
Rep. American Chemistry Council

John R. Kovacik, UL LLC, IL [RT]

Alan Manche, Schneider Electric, KY [M]

Richard P. Owen, Oakdale, MN [E]
Rep. International Association of Electrical Inspectors

James F. Pierce, Intertek Testing Services, OR [RT]

Vincent J. Saporita, Eaton's Bussmann Business, MO [M]
Rep. National Electrical Manufacturers Association

Alternates

Lawrence S. Ayer, Biz Com Electric, Inc., OH [IM]
(Alt. to David L. Hittinger)

Roland E. Deike, Jr., CenterPoint Energy, Inc., TX [UT]
(Voting Alt.)

James T. Dollard, Jr., IBEW Local Union 98, PA [L]
(Alt. to Palmer L. Hickman)

Stanley J. Folz, Morse Electric Company, NV [IM]
(Alt. to Michael J. Johnston)

Ernest J. Gallo, Telcordia Technologies (Ericsson), NJ [UT]
(Alt. to James E. Brunssen)

Robert A. McCullough, Tuckerton, NJ [E]
(Alt. to Richard P. Owen)

Mark C. Ode, UL LLC, AZ [RT]
(Alt. to John R. Kovacik)

Christine T. Porter, Intertek Testing Services, WA [RT]
(Alt. to James F. Pierce)

George A. Straniero, AFC Cable Systems, Inc., NJ [M]
(Alt. to Vincent J. Saporita)

Nonvoting

Timothy J. Pope, Canadian Standards Association, Canada [SE]
Rep. CSA/Canadian Electrical Code Committee

William R. Drake, Fairfield, CA [M]
(Member Emeritus)

Mark W. Earley, NFPA Staff Liaison

D. Harold Ware, Libra Electric Company, OK [IM]
(Member Emeritus)

Reprinted with permission from NFPA 70-2017, *National Electrical Code*®, Copyright© 2016, National Fire Protection Association, Quincy, MA 02169. This reprinted material is not the complete and official position of the NFPA on the referenced subject, which is represented only by the standard in its entirety.

Figure A-4. The Correlating Committee is the watchdog that ensures there are no conflicting actions between CMPs.

Figure A-5 **2017 *NEC Code*-Making Panels**

***NEC CODE*-MAKING PANEL**	**ARTICLES, ANNEX AND CHAPTER 9 MATERIAL WITHIN THE SCOPE OF THE *CODE*-MAKING PANEL**
1	90, 100, 110, Chapter 9, Table 10, Annex A, Annex H, Annex I, Annex J
2	210, 215, 220, Annex D Examples D1 through D6
3	300, 590, 720, 725, 727, 728, 760, Chapter 9, Tables 11(A) and (B), Tables 12(A) and (B)
4	225, 230, 690, 692, 694, 705, 710
5	200, 250, 280, 285
6	310, 400, 402, Chapter 9 Tables 5 through 9, and Annex B
7	320, 322, 324, 326, 328, 330, 332, 334, 336, 338, 340, 382, 394, 396, 398, 399
8	342, 344, 348, 350, 352, 353, 354, 355, 356, 358, 360, 362, 366, 368, 370, 372, 374, 376, 378, 380, 384, 386, 388, 390, 392, Chapter 9, Tables 1 through 4, Example D13, and Annex C
9	312, 314, 404, 408, 450, 490
10	240
11	409, 430, 440, 460, 470, Annex D, Example D8
12	610, 620, 625, 626, 630, 640, 645, 647, 650, 660, 665, 668, 669, 670, 685, Annex D, Examples D9 and D10
13	445, 455, 480, 695, 700, 701, 702, 706, 708, 712, 750, Annex F, and Annex G
14	500, 501, 502, 503, 504, 505, 506, 510, 511, 513, 514, 515, 516
15	517, 518, 520, 522, 525, 530, 540
16	770, 800, 810, 820, 830, 840
17	422, 424, 425, 426, 427, 680, 682
18	393, 406, 410, 411, 600, 605
19	545, 547, 550, 551, 552, 553, 555, 604, 675, and Annex D, Examples D11 and D12

Figure A-5. There are 19 CMPs for the NEC.

Figure A-6 **2017 *NEC* CMP No. 3 Members**

CODE-MAKING PANEL NO. 3

Articles 300, 590, 720, 725, 727, 728, 760, Chapter 9, Tables 11(A) and (B), and Tables 12(A) and (B)

Paul J. Casparro, *Chair*
Scranton Electricians JATC, PA [L]
Rep. International Brotherhood of Electrical Workers

Douglas P. Bassett, XFinity Home, FL [IM]
Rep. Electronic Security Association
(VL to 720, 725, 727, 760)

Larry G. Brewer, Intertek Testing Services, NC [RT]

William A. Brunner, Main Electric Construction Inc., ND [IM]
Rep. National Electrical Contractors Association

Steven D. Burlison, Progress Energy, FL [UT]
Rep. Electric Light & Power Group/EEI

Shane M. Clary, Bay Alarm Company, CA [M]
Rep. Automatic Fire Alarm Association, Inc.

Adam D. Corbin, Corbin Electrical Services, Inc., NJ [IM]
Rep. Independent Electrical Contractors, Inc.

Les Easter, Atkore International, IL [M]
Rep. National Electrical Manufacturers Association

Ray R. Keden, Pentair-ERICO, CA [M]
Rep. Building Industry Consulting Services International

T. David Mills, T. David Mills Associates, LLC, GA [U]
Rep. Institute of Electrical & Electronics Engineers, Inc.

Steven J. Owen, Steven J. Owen, Inc., AL [IM]
Rep. Associated Builders & Contractors

David A. Pace, Olin Corporation, AL [U]
Rep. American Chemistry Council

Susan Newman Scearce, City of Humboldt, TN, TN [E]
Rep. International Association of Electrical Inspectors

John E. Sleights, Travelers Insurance Company, CT [I]

Susan L. Stene, UL LLC, CA [RT]

Alternates

Richard S. Anderson, RTKL Associates Inc., VA [M]
(Alt. to Ray R. Keden)

Jorge L. Arocha, Florida Power & Light, FL [UT]
(Alt. to Steven D. Burlison)

Sanford E. Egesdal, Egesdal Associates PLC, MN [M]
(Alt. to Shane M. Clary)

Michael J. Farrell III, Lucas County Building Regulation, MI [L]
(Alt. to Paul J. Casparro)

Danny Liggett, The DuPont Company, Inc., TX [U]
(Alt. to David A. Pace)

Mark C. Ode, UL LLC, AZ [RT]
(Alt. to Susan L. Stene)

Dmitriy V. Plotnikov, Intertek Testing Services, NJ [RT]
(Alt. to Larry G. Brewer)

Rick D. Sheets, DIRECTV, TX [IM]
(VL to 720, 725, 727, 760)
(Alt. to Douglas P. Bassett)

George A. Straniero, AFC Cable Systems, Inc., NJ [M]
(Alt. to Les Easter)

Joseph J. Wages, Jr., International Association of Electrical Inspectors, TX [E]
(Alt. to Susan Newman Scearce)

Nonvoting

Edward C. Lawry, Oregon, WI [E]
(Member Emeritus)

Reprinted with permission from NFPA 70-2017, *National Electrical Code*®, Copyright© 2016, National Fire Protection Association, Quincy, MA 02169. This reprinted material is not the complete and official position of the NFPA on the referenced subject, which is represented only by the standard in its entirety.

Figure A-6. CMP No. 3 for the NEC *has an IBEW chairman from the Scranton Electricians JATC, PA.*

information such as which organization the applicant would represent. The committee membership is made up of highly qualified personnel with public safety as a high priority.

Classification

The CMPs, also known as the TCs, are volunteers. A listing of the scope for each CMP and the committee list appear in the front of the *NEC*. In the committee list, TC member names are followed by their employer names and identification letter(s). The identification letters appear in brackets, such as [L] for Labor. Most organizations represented have a principal member and an alternate member. Committee membership classification is

Figure A-7 Committee Member Classifications

M	Manufacturers: makers of products affected by the *NEC*
U	Users: users of the *NEC*
I/M	Installers/Maintainers: installers/maintainers of systems covered by the *NEC*
L	Labor: those concerned with safety in the workplace
R/T	Research/Testing Labs: independent organizations developing/reinforcing standards
E	Enforcing Authority: inspectors, enforcers of the *NEC*
I	Insurance: insurance companies, bureaus, or agencies
C	Consumers: purchasers of products/systems not included in "U," users
SE	Special Experts: providers of special expertise, not applicable to other classifications
UT	Utilities: installers/maintainers of systems not covered by the *NEC*

Figure A-7. CMP members come from a wide background of expertise, such as manufacturers and research labs.

part of the balancing process of each panel. **See Figure A-7**.

Representation

Many organizations take part in the *NEC* process and provide representation for the membership classification that applies to their organization. For example, the members of the TCs with the classification "E," for Enforcing Authority, are representatives of the IAEI, International Association of Electrical Inspectors. **See Figures A-8 and A-9**.

Figure A-8 NFPA Electrical Engineering Division Technical Staff

NFPA Electrical Engineering Division Technical Staff

William Burke, Division Manager
Mark W. Earley, Chief Electrical Engineer
Mark Cloutier, Senior Electrical Engineer
Christopher Coache, Senior Electrical Engineer
Carol Henderson, Technical Administrator
Richard J. Roux, Senior Electrical Specialist
Kimberly L. Shea, Project Administrator
Derek Vigstol, Senior Electrical Specialist
Mary Warren, Technical Administrator

Reprinted with permission from NFPA 70-2017, *National Electrical Code*®, Copyright© 2016, National Fire Protection Association, Quincy, MA 02169. This reprinted material is not the complete and official position of the NFPA on the referenced subject, which is represented only by the standard in its entirety.

Figure A-8. The NFPA staff consist of technical, support, and editorial personnel.

Figure A-9

Organizations Represented	
Air-Conditioning, Heating, & Refrigeration Institute	Insulated Cable Engineers Association Incorporated
Alliance for Telecommunications Industry Solutions	International Association of Electrical Inspectors
Alliance of Motion Picture and Television Producers	International Alliance of Theatrical Stage Employees
The Aluminum Association Incorporated	International Brotherhood of Electrical Workers
American Chemistry Council	International Electrical Testing Association Incorporated
American Iron and Steel Institute	International Sign Association
American Institute of Organ Builders	National Association of Homebuilders
American Lighting Association	National Association of RV Parks and Campgrounds
American Petroleum Institute	Recreational Vehicle Industry Association
American Society of Agricultural & Biological Engineers	National Cable & Telecommunications Association
American Society for Healthcare Engineering	National Electrical Contractors Association
American Wind Energy Association	National Electrical Manufacturers Association
Associated Builders and Contractors	National Elevator Industry Incorporated
Association of Higher Education Facilities Officers	Outdoor Amusement Business Association Incorporated
Association of Pool & Spa Professionals	Power Tool Institute Incorporated
Automatic Fire Alarm Association	Recreational Vehicle Industry Association
Building Industry Consulting Service International	Satellite Broadcasting & Communications Association
CSA/Canadian Electrical Code Committee	Society of Automotive Engineers - Hybrid Committee
Copper Development Association Incorporated	Society of the Plastics Industry Incorporated
Electric Light & Power Group/EEI	Solar Energy Industries Association
Electronic Security Association	TC on Airport Facilities
Grain Elevator and Processing Society	TC on Electrical Systems
Illuminating Engineering Society of North America	Telecommunications Industry Association
Independent Electrical Contractors	Transportation Electrification Committee
Information Technology Industry Council	U.S. Institute for Theatre Technology
Institute of Electrical & Electronics Engineers	The Vinyl Institute
Instrumentation, Systems, & Automation Society	

Figure A-9. Many interested organizations are members of CMPs.

Summary

The *NEC* revision process is open to all members of the public who wish to take part by submitting proposed changes. The many organizations that participate in the *NEC Code*-making process help to build a true consensus code. Since the *NEC* will change every three years, memorizing requirements or sections may be a wasted effort. The *Codeology* method, however, will not change. Applying the *Codeology* method will lead to the quick and accurate location of necessary information in the *NEC* today and in all future and past versions of the *Code*.

A Brief Guide for New Users of the *NEC*

Introduction

A qualified person working in the electrical industry must have the proper skills, knowledge, and aptitude to design, install, and maintain safe electrical installations. The *electrical training ALLIANCE* provides standardized electrical training to accomplish this goal. The *National Electrical Code (NEC)*, published by the National Fire Protection Association (NFPA), is a critical reference tool for anyone working in the electrical industry. Users of the *NEC* can benefit from an organized system designed to address the layout and content of the *Code*. The *electrical training ALLIANCE*'s *Applied Codeology: Navigating the NEC* textbook method provides the electrical professional a systematic way to find information in the vast catalog of articles contained in the *NEC*.

Appendix B

Table of Contents

ROLE OF THE *ELECTRICAL TRAINING ALLIANCE*

The *electrical training ALLIANCE* provides the best standardized training programs available for the electrical industry. Thousands of the best apprentice and journeyman Electrical Workers in the industry are trained and update their skills each year through the *electrical training ALLIANCE* curriculum. These educational training programs are provided by local apprenticeship training committees affiliated with the International Brotherhood of Electrical Workers (IBEW) and the National Electrical Contractors Association (NECA). NECA contractors utilize these best-trained Electrical Workers for installing and servicing the electrical power and distribution needs of the country.

The electrical industry relies on qualified persons and craftspeople to install, repair, and maintain electrical systems. A *qualified person* is one with skills and knowledge relating to the construction and operation of electrical equipment and installation. This means a qualified Electrical Worker has completed both on-the-job and classroom training and in some cases has demonstrated competency by passing a qualifying examination through a certified organization or state licensing process.

The worker must not only have basic skills and knowledge of construction and installation of electrical systems and equipment. A completely trained and qualified worker also understands the second, equally important part of being a qualified person, namely, receiving safety training to recognize and avoid hazards associated with the uses of electricity. This second qualification is provided through worker OSHA training, particularly through knowledge of *NFPA 70E: Standard for Electrical Safety in the Workplace*. The *electrical training ALLIANCE* provides an extensive training curriculum for worker qualification in these vitally important safety areas. **See Figure B-1**.

Figure B-1 *electrical training ALLIANCE* Curriculum

Electrical Safety Related Work Practices

Fourth Edition

Based on *NFPA 70E®* 2015 Edition

NJATC's Course on OSHA Outreach Training Volume 2

Figure B-1. *The* electrical training ALLIANCE *curriculum provides study materials for qualified persons relating to on-the-job safety from OSHA and the most current editions of* NFPA Standards *and* Code.

WHAT IS THE *CODEOLOGY* METHOD?

One of the more challenging aspects of learning the electrical trade is mastering the *National Electrical Code*. **Article 90.1(A)** informs the user that the *NEC* is not a design manual or a book used as an instruction manual for untrained persons. Using the *Code* can be difficult and frustrating for those who do not understand its layout and content. Job-site productivity and profitability can be severely compromised when an inexperienced designer or worker unknowingly violates an *NEC* rule. Using the *Codeology* method helps the designer, foreman, worker, and even a candidate for a state qualifying exam to use the *Code* proficiently and to find applicable *Code* references and answers to questions quickly.

The process of mastering how to find information in the *NEC* takes time, as it requires learning about both content and application. As apprentices gain more experience in the electrical industry, they find the *NEC* easier to use.

The *Codeology* method employs the *NEC*'s index, clues, and keywords, the four building blocks, as well as the concepts of plan, build, and use to find applicable *Code* sections. There are many examples and opportunities provided to practice and hone *Codeology* skills. Students should be persistent and exercise patience, looking up every reference necessary to answer the provided questions.

NEC DEFINITIONS

For example, in the *Manual of Style for NFPA Technical Committee Documents,* the term *consensus* is defined as a substantial agreement to a proposed change of language in a code or standard that has been reached by the affected interest categories. Substantial agreement means much more than a simple majority; two-thirds of the Technical Committee members must agree, but the vote to adopt a proposal is not necessarily by unanimity. Consensus requires that all views and objections be considered by the Technical Committee and that a concerted effort is made toward resolving conflicts or differing opinions. All committee decisions are reviewed by the Standards Council, which bases its judgment on a proposed change when a consensus has been achieved.

The *Manual of Style for NFPA Technical Committee Documents* defines a code as a standard that has an extensive gathering of requirements covering broad subject matter. A code is suitable for adoption into law independently of other codes and standards. The decision to designate a standard as a code is based on the size and scope of the document, its intended use, whether it can be adopted as law, and whether it is suitable for enforcement and has administration provisions.

A *standard* is defined as a document where the main text contains mandatory provisions using the word "shall" to indicate requirements and is written in a form generally suitable for mandatory reference by another code or standard. For example, *NEC* **250.106** requires lightning protection systems to be bonded to the building or structure grounding electrode system, and the Informational Note refers the user to *NFPA 780-2014: Standard for the Installation of Lightning Protection Systems*. If a standard contains non-mandatory provisions, they shall be located in an appendix or annex, footnote, or fine-print Informational Note and not considered part of the requirements of the standard.

TECH FACT

An example of a guide is *NFPA 921: Guide for Fire and Explosion Investigations.* This document is designed to assist individuals who are charged with the responsibility of investigating and analyzing fire and explosion incidents and rendering opinions as to the origin, cause, responsibility, or prevention of such incidents.

The NFPA also publishes guides and recommended practice documents. Guides are advisory or informative in nature, containing only non-mandatory provisions. A guide may contain mandatory language, but the document as a whole is not suitable for adoption into law.

A recommended practice published by NFPA is a document that is similar in content and structure to a code or standard but contains only non-mandatory provisions using the word "should" to indicate recommendations in the body of the text.

ENFORCEMENT

The *NEC* is purely advisory as far as the NFPA is concerned. It is made available for both private and public interests regarding life safety and protection of property from hazards arising from the use of electricity. The *NEC* can be applied to both law and regulatory purposes in all 50 states. It is used for private self-regulation and standardization activities such as insurance underwriting, building and facilities construction and management, and product testing and certification.

Governing bodies have the authority to enforce the provisions of the *NEC* when

TECH FACT

An example of a recommended practice is *NFPA 70B: Recommended Practice for Electrical Equipment Maintenance. NFPA 70B* provides guidance necessary to develop and carry out an effective preventative maintenance (EPM) program for all types of equipment and assemblies.

it is adopted into law by a particular jurisdiction (**Section 90.4**). The authority having jurisdiction (AHJ) is defined in **Article 100** as an organization or person responsible for enforcing the code or standard and issuing approvals of equipment and installations covered by the rules of the *NEC*.

- **Authority Having Jurisdiction (AHJ).** An organization, office, or individual responsible for enforcing the requirements of a code or standard, or for approving equipment, materials, an installation, or a procedure.

Note

The phrase "authority having jurisdiction," or its acronym, AHJ, is used in NFPA documents in a broad manner, since jurisdictions and approval agencies vary, as do their responsibilities. Where public safety is primary, the authority having jurisdiction may be a federal, state, local, or other regional department or individual such as a fire chief; fire marshal; chief of a fire prevention bureau, labor department, or health department; building official; electrical inspector; or others having statutory authority. For insurance purposes, an insurance inspection department, rating bureau, or other insurance company representative may be the authority having jurisdiction. In many circumstances, the property owner or his or her designated agent assumes the role of the authority having jurisdiction. At government installations, the commanding officer or departmental official may be the authority having jurisdiction.

The AHJ is also responsible for interpreting the requirements and granting special permission as provided in some of the rules.

To most electricians, the terms *authority having jurisdiction* or *AHJ* means a building or electrical inspector. The terms are used in the *NEC* in a broad manner as described in the Informational Note to the definition for authority having jurisdiction in **Article 100**. Federal, state, and local government agencies, fire departments or fire chiefs, labor and health departments, as well as building officials and electrical inspectors all have statutory responsibility for code enforcement.

Most governmental bodies that adopt the *NEC* require a complete process of consideration and inspections for plans to be completed. Most require inspections of open walls before building finishes cover them. This means the AHJ will visit a job site during the course of construction to ensure that the building of electrical installations complies with the *NEC*. The AHJ should be treated with respect at all times and be afforded the opportunity to inspect the entire electrical installation as necessary. The inspection requirement can vary with jurisdictions, so the inspection and approval procedures required in the jurisdiction where the project is being built should always be verified.

Summary

The *electrical training ALLIANCE* provides apprentice and journeyman training for thousands of Electrical Workers every year. *Applied Codeology Navigating the NEC* uses clues, keywords, and the four basic building blocks to improve *Code* book skills. A qualified person has skills and knowledge relating to the construction and operation of electrical equipment and recognizes and avoids the hazards involved. Important *NEC* definitions are provided in the *Manual of Style for NFPA Technical Committee Documents.* While *NEC* enforcement is purely advisory, governing bodies have the authority to enforce *NEC* provisions when they are adopted into law. The AHJ enforces the *NEC* where adopted in a jurisdiction.

Index